Synthese Library

Studies in Epistemology, Logic, Methodology, and Philosophy of Science

Volume 471

The aim of *Synthese Library* is to provide a forum for the best current work in the methodology and philosophy of science and in epistemology, all broadly understood. A wide variety of different approaches have traditionally been represented in the Library, and every effort is made to maintain this variety, not for its own sake, but because we believe that there are many fruitful and illuminating approaches to the philosophy of science and related disciplines.

Special attention is paid to methodological studies which illustrate the interplay of empirical and philosophical viewpoints and to contributions to the formal (logical, set-theoretical, mathematical, information-theoretical, decision-theoretical, etc.) methodology of empirical sciences. Likewise, the applications of logical methods to epistemology as well as philosophically and methodologically relevant studies in logic are strongly encouraged. The emphasis on logic will be tempered by interest in the psychological, historical, and sociological aspects of science. In addition to monographs *Synthese Library* publishes thematically unified anthologies and edited volumes with a well-defined topical focus inside the aim and scope of the book series. The contributions in the volumes are expected to be focused and structurally organized in accordance with the central theme(s), and should be tied together by an extensive editorial introduction or set of introductions if the volume is divided into parts. An extensive bibliography and index are mandatory.

Yang Liu • Stephan Hartmann • Huw Price
Editors

Decision Theory and the Future of AI

Previously published in *Synthese* "Special Issue: Decision Theory and the Future of AI" Volume 198, supplement issue 27, November 2021

Springer

Editors
Yang Liu
Faculty of Philosophy and CFI
University of Cambridge
Cambridge, UK

Stephan Hartmann
Munich Center Math Philo
Ludwig Maximilians-Universität München
München, Bayern, Germany

Huw Price
Faculty of Philosophy
University of Cambridge
Cambridge, UK

Spinoff from journal: "Synthese" Volume 198, supplement issue 27, November 2021

ISSN 0166-6991
Synthese Library
ISBN 978-3-031-19281-4

This Springer imprint is published by the registered company Springer Nature Switzerland AG
The registered company address is: Gewerbestrasse 11, 6330 Cham, Switzerland

Contents

Synthese (2021) 198 (Suppl 27):S6413–S6414
https://doi.org/10.1007/s11229-021-03316-z

S.I. : DECTHEORY&FUTOFAI

Editorial to "Decision theory and the future of AI"

Yang Liu[1] · Stephan Hartmann[2] · Huw Price[3]

Published online: 12 September 2021

In the long run, the development of artificial intelligence (AI) is likely to be one of the biggest technological revolutions in human history. Even in the short run it will present tremendous challenges as well as tremendous opportunities. The more we do now to think through these complex challenges and opportunities, the better the prospects for the kind of outcomes we all hope for, for ourselves, our children, and our planet.

Thinking through these challenges needs a new kind of interdisciplinary research community. Many sources of expertise and insight are likely to be relevant, and this community needs to be very well-connected in several dimensions—'horizontally' between academic disciplines, 'vertically' to the policy and technology worlds, and of course geographically. AI is a global technology, and many of the challenges and opportunities of AI will be global in nature. Accordingly, getting AI right is not just an engineering challenge, but also a challenge for many other societal and academic sectors, including the humanities. Put another way, there is an engineering challenge of a 'sociological' kind, about how best to foster the necessary research community.

The field of decision theory is ideally placed to make contributions here, at several levels. AI innovations, including techniques from machine learning, are increasingly used to make *decisions* with significant social and ethical consequences, ranging from determining the news feeds on social media to making sentencing and parole recommendations in the criminal justice system. Decision theory provides and studies the standards by which such decisions are evaluated and improved. What is a rational decision? How can we train machines to make rational decisions? What is the relationship between human decision-making and machine decision-making? How can one make machine decision-making transparent (i.e. understandable to a human agent)? Which role does cognitive science play in these developments?

Chapter 1 was originally published as Liu, Y., Hartmann, S. & Price, H. Synthese (2021) 198 (Suppl 27): S6413–S6414. https://doi.org/10.1007/s11229-021-03316-z.

✉ Yang Liu
yl587@cam.ac.uk

1 Faculty of Philosophy and Leverhulme Centre for the Future of Intelligence (CFI), University of Cambridge, Level 1, 16 Mill Lane, Cambridge CB2 1SB, UK

2 Munich Center for Mathematical Philosophy of Science, LMU Munich, Geschwister-Scholl-Platz 1, 80539 Munich, Germany

3 Leverhulme Centre for the Future of Intelligence (CFI), University of Cambridge, Level 1, 16 Mill Lane, Cambridge CB2 1SB, UK

Perhaps even more importantly, the field of decision theory itself is highly interdisciplinary, with a strong presence in disciplines such as philosophy, mathematical logic, economics, psychology, and cognitive science, amongst others. In addition, of course, it has foundational links to computer science and machine learning. So it is ideally placed to contribute to the sociological project, offering fertile ground in which to foster the interdisciplinary community needed for the challenges of AI, short term and long term.

This special issue stems from a conference series established with these goals in mind. *Decision Theory and the Future of AI* began in 2017 as a collaboration between the Leverhulme Centre for the Future of Intelligence (CFI) and the Centre for the Study of Existential Risk (CSER) at Cambridge, and Munich Center for Mathematical Philosophy (MCMP) at LMU Munich. The first two conferences were held at Trinity College, Cambridge in 2017 and LMU Munich in 2018. The first meeting outside Europe was held at ANU, Canberra, in 2019, in conjunction with ANU's Humanising Machine Intelligence project. A fourth conference was planned at PKU, Beijing, in 2020, before Covid intervened. We will be back!

Several of the papers in this special issue were presented at one of these conferences, while others were submitted in response to an open call for papers. The range of topics, and even more so the range of authors and their home disciplines and affiliations, is a tribute to the richness of the territory, both in intellectual and in community-building terms.

Acknowledgements We would like to thank the Cambridge-LMU Strategic partnership for support for the conference series, and our institutions (CFI, CSER, and the Faculty of Philosophy at Cambridge, and MCMP at LMU Munich) for administrative support. We also thank the authors and referees of the papers for all their work. Yang Liu and Huw Price gratefully acknowledge the support of the Leverhulme Trust, and of a grant from Templeton World Charity Foundation (TWCF0128); the opinions expressed in this publication are those of the authors and do not necessarily reflect the views of TWCF.

Synthese (2021) 198 (Suppl 27):S6415–S6434
https://doi.org/10.1007/s11229-019-02272-z

S.I.: DECTHEORY&FUTOFAI

A classification of Newcomb problems and decision theories

Kenny Easwaran[1]

Received: 24 March 2018 / Accepted: 28 May 2019 / Published online: 30 May 2019

Abstract
Newcomb-like problems are classified by the payoff table of their act-state pairs, and the causal structure that gives rise to the act-state correlation. Decision theories are classified by the one or more points of intervention whose causal role is taken to be relevant to rationality in various problems. Some decision theories suggest an inherent conflict between different notions of rationality that are all relevant. Some issues with causal modeling raise problems for decision theories in the contexts where Newcomb problems arise.

Keywords Newcomb problem · Act-state dependence · Causal modeling · Decision theory · Rationality

1 Introduction

Consider the following two decision problems.

Example 1: The Newcomb Problem

You walk into a booth at a carnival and find a strange game being offered. There is a mad scientist, who offers you the option of taking just the contents of an opaque box, or the contents plus another box containing a thousand dollars. But as you walked in, her machines scanned your brain and body, and using the laws of physics and neurology, her computers predicted what you were going to choose. If they predicted that you would choose just the contents of the opaque box, she put a million dollars in it. But if they predicted that you would take both boxes, or if they couldn't predict what you would do, then she put nothing in. While you are deliberating, you learn that her predictions in the past have been moderately reliable, so that a significant majority (though not all) of her predictions of each type have been correct, conditioning either on her prediction

Chapter 2 was originally published as Easwaran, K. Synthese (2021) 198 (Suppl 27): S6415–S6434. https://doi.org/10.1007/s11229-019-02272-z.

✉ Kenny Easwaran
easwaran@tamu.edu

1 Texas A&M University, College Station, USA

or on the choice made by past participants. Should you just take the contents of the one opaque box, or take both boxes?

Example 2: The Smoking Lesion

Scientists discover that the statistician R.A. Fisher's work on behalf of the tobacco companies was actually correct. Although there is a strong correlation between smoking and lung cancer, it turns out that this is not because of any mechanism by means of which the nicotine, tar, or anything else affects human bodies. Rather, it is because of a previously unknown physiological condition that both has a tendency to cause cancerous growths in the lungs (and is the primary source of these growths), and seems to behaviorally lead to smoking. (This condition is caused by environmental factors that happen to have been cleaned up greatly in North America over the past few decades, but are still quite high in developing countries). You have had some cigarettes in the past, and have moderately enjoyed them. Should you take up smoking?

Classic "evidential decision theory" notes that in each case, the fact that an agent takes one of the options (taking both boxes, smoking) is good evidence that she will receive quite a bad outcome (failing to get her million dollars, cancer), and thus says that a rational agent who knows the setup will choose the other option. Classic "causal decision theory" notes that in each case, at the moment of decision, the features of the world that give rise to the bad outcome (prediction, physiological condition) are already fixed and outside of the agent's control, and instead focuses on the fact that the option that is correlated with this outcome will definitely cause a moderately good outcome (getting a bonus thousand, the enjoyment of a cigarette), and thus says that a rational agent who knows the setup will choose this option. Many people have expressed intuitions that cross-cut these cases, and thus some theorists have gone to great lengths to try to define versions of decision theory that recommend taking only one box in The Newcomb Problem, but smoking in The Smoking Lesion. (See, among others, the "ratifiability" of Jeffrey (1981), the "tickle defense" of Eells (1982), the attack on causal decision theory by Price (1986), the "cohesive decision theory" of Meacham (2010), and the "timeless decision theory" of Yudkowsky (2010)).

This paper does not seek to adjudicate between these theories, but instead attempts to give a structure for classifying problems of the relevant sort, and to use this structure to give some insight into the classification of possible decision theories for dealing with them. I classify the problems by means of a payoff table (which these two examples happen to share), together with a causal model explaining the correlation between the row and column of the payoff table (on which these two problems differ). I classify the decision theories by citing which particular variable in the causal model is the one that should be conditioned (evidential decision theory) or intervened on (the act itself for causal decision theory, and the decision procedure for timeless decision theory) in order to generate the probabilities used to evaluate the rows of the payoff table. (I consider decision theories as evaluative theories that classify certain acts as rational or irrational, rather than as procedures that agents themselves ought to use to generate their decisions).

These two classification systems then suggest some ways of generating new cases of the relevant sort (by combining some particular payoff table with a causal model), and new theories as well (by finding a different variable to condition or intervene on). I consider some motivations for thinking that one of these theories might be the correct account of decision-theoretic rationality, or whether all might be correct accounts of different parts of it. I close with discussion of some interactions of the concepts of free will, determinism, and rationality that are suggested by these classifications, though there are no definitive claims about these issues.

2 Classifying decision problems

One of the puzzling features of these cases is the fact that the "states of the world" are not probabilistically independent of the agent's choice. Thus, these decisions can't be properly analyzed by the methods proposed by Savage (1954) without having vastly more information than is available. (Savage requires a formulation of the problem where the states identified really are independent of the act chosen, which *might* involve "states" like "the predictor predicts correctly", or might involve some more subtle analysis. A proper Savage-style formulation of Newcomb-type problems is a major problem that is beyond the scope of this paper). Decision theorists dealing with Newcomb-type problems have more often tried to deal with them by giving some way to evaluate acts that are not always assumed to be independent of the "states".

For simplicity, I will focus on decision problems that are naturally described in terms of two possible acts and two possible states, which have some sort of correlation with each other. My proposed classification of these problems consists of two parts. First, I will characterize the value of the possible outcomes to the agent by giving a payoff table. Second, I will describe the structure of common cause relations giving rise to the correlation between the states and the acts. Although these parts leave out most of the real-world texture and background of the problems, I suspect that they are largely adequate for the purposes of decision theory. Importantly, although The Newcomb Problem and The Smoking Lesion have the same payoff table of possible outcomes, they will have different common cause structures, and thus will not be classified the same on my account, the way they are often treated in the literature. This will open space for decision theories that recommend one-boxing in The Newcomb Problem and smoking in The Smoking Lesion, even though other decision theories ignore the differences between the common cause structures.

2.1 Payoff tables

I represent the values of the possible outcomes for an agent in a 2×2 matrix like $\left(\begin{smallmatrix} a & b \\ c & d \end{smallmatrix}\right)$, which I call a "payoff table". Each row of the payoff table corresponds to one of the possible actions the agent can take and each column corresponds to one of the states. The contents of each cell then represents the value of the outcome that would be achieved by the corresponding combination of act and state. In most cases, the precise value of the outcome isn't relevant, so we will only be concerned with the ordinal structure of the value of these four outcomes. Thus, I will generally represent the values in the payoff table with the numerals 1–4, with 1 representing the best outcome and 4 the worst. When relevant, I will supplement this with some more detailed comparisons—for instance, $1 > 2 \gg 3 \sim 4$ will represent the fact that for the analysis of the relevant decision problem, it matters that the best option is better than the second best, which is much better than the third and fourth, which are approximately equal. When merely ordinal information is relevant, there are 24 possible payoff tables, though there are more if these magnitudes of comparison are considered.

Since switching the two rows just corresponds to relabeling the actions the agent can take, and switching the two columns just corresponds to relabeling the states, each decision problem has four possible payoff tables. I will cut this to two by following the convention that the act corresponding to the top row is positively probabilistically correlated with the state corresponding to the left column, and the act corresponding to the bottom row is positively correlated with the state corresponding to the right column. (When we look at the structure of the correlations in Sect. 2.2, it will be clear that there are further complications that I am ignoring right now). To remove this last degree of freedom, I will choose the representation where the upper left correlated outcome is strictly better than the lower right one. (If the upper left and lower right outcomes are equal, then I will ask that the upper right outcome is strictly better than the lower left one. If those are equal too, then both representations will in fact be identical). In addition to giving each decision problem a unique payoff table (at least, when the two correlated outcomes have distinct values), this eliminates half of the possible payoff tables, so that there are only 12 when merely ordinal information is represented.

The first thing to note is that both The Newcomb Problem and The Smoking Lesion are represented by the payoff table $\left(\begin{smallmatrix}2 & 4\\ 1 & 3\end{smallmatrix}\right)$, with $1 > 2 \gg 3 > 4$. Each outcome of the lower action (taking two boxes, smoking) is slightly better than the outcome of the upper action (taking one box, not smoking), but the outcomes of the correlated left state (million dollars, no cancer) are much better than the outcomes of the right state (empty box, cancer). Several other puzzles that are prominent in the literature can be represented with the same payoff table.

Example 3: The Toxin Puzzle

> (Kavka 1983, p. 33–34) You have just been approached by an eccentric billionaire who has offered you the following deal. He places before you a vial of toxin that, if you drink it, will make you painfully ill for a day, but will not threaten your life or have any lasting effects. (Your spouse, a crack biochemist, confirms the properties of the toxin). The billionaire will pay you one million dollars tomorrow morning if, at midnight tonight, you *intend* to drink the toxin tomorrow afternoon. He emphasizes that you need not drink the toxin to receive the money; in fact, the money will already be in your bank account hours before the time for drinking it arrives, if you succeed. (This is confirmed by your daughter, a lawyer, after she examines the legal and financial documents that the billionaire has signed). All you have to do is sign the agreement and then intend at midnight

> tonight to drink the stuff tomorrow afternoon. You are perfectly free to change your mind after receiving the money and not drink the toxin. (The presence or absence of the intention is to be determined by the latest 'mind-reading' brain scanner and computing device designed by the great Doctor X. As a cognitive scientist, materialist, and faithful former student of Doctor X, you have no doubt that the machine will correctly detect the presence or absence of the relevant intention).

Ignoring the central issue that Kavka uses this puzzle to discuss (namely, the question of whether it is even possible to form the relevant intention, knowing that your future self will have some incentive to go back on it), the value structure is quite similar. Regardless of whether or not you formed the intention tonight at midnight, not drinking the toxin tomorrow will be somewhat better than drinking the toxin. However, regardless of whether or not you drink the toxin, having formed the intention tonight results in a substantially better outcome than not having formed the intention. There is presumably some sort of correlation between having formed the intention and drinking the toxin, and it's better to have formed the intention and drink the toxin than to do neither.

Example 4: Parfit's Hitchhiker

> (Parfit 1984, p. 7) Suppose that I am driving at midnight through some desert. My car breaks down. You are a stranger and the only other driver near. I manage to stop you, and I offer you a great reward if you rescue me. I cannot reward you now, but I promise to do so when we reach my home. Suppose next that I am transparent, unable to deceive others. I cannot lie convincingly. Either a blush, or my tone of voice, always gives me away.

Here again we have an issue for planning. If I sincerely plan now to give you the reward when I get home, then regardless of whether or not I actually do give you the reward, I will get the major benefit of a ride out of the desert. However, actually giving you the reward is a cost to me, and it is correlated (in some sense to be discussed later) with the sincere plan. (In this case, if I don't plan to give you the reward, I may not even get the option to give it, because I'm stuck in the desert, so we have $3 \sim 4$ rather than $3 > 4$).

Example 5: Prisoner's Dilemma with a Twin

> You and your twin sibling have robbed a bank. However, you were caught and are now being held in separate cells. The police have enough evidence currently to convict each of you of conspiracy. However, they don't have enough evidence to convict either of you of actually robbing the bank. The police say that if you testify against your sibling (giving them enough evidence to convict), they'll drop your conspiracy charge. They're also making the same offer to your sibling in the other cell. Conspiracy carries a sentence of a year in jail, but bank robbery carries a sentence of five years.

In this case, regardless of whether or not your sibling testifies, if you testify, you'll spend one less year in jail. However, because your twin is very much like you, your testifying is strongly correlated with your twin testifying, which would add five years to your jail sentence whether you testify or not.

Further analysis of the similarities and differences among these decision puzzles will wait until Sect. 2.2.

The distinct, but related, payoff table $\left(\begin{smallmatrix}3 & 2\\ 1 & 4\end{smallmatrix}\right)$ also corresponds to a few prominent puzzles in the literature.

Example 6: Death in Damascus

(Gibbard and Harper 1978, pp. 185–186) Death works from an appointment book which states time and place; a person dies if and only if the book correctly states in what city he will be at the stated time. The book is made up weeks in advance on the basis of highly reliable predictions. An appointment on the next day has been inscribed for him. Suppose, on this basis, the man would take his being in Damascus the next day as strong evidence that his appointment with death is in Damascus, and would take his being in Aleppo the next day as strong evidence that his appointment is in Aleppo. Two acts are open to him: A, go to Aleppo, and D, stay in Damascus. There are two possibilities: SA, death will seek him in Aleppo, and SD, death will seek him in Damascus. He knows that death will find him if and only if death looks for him in the right city.

In this case, the rows correspond to the actions of going to Aleppo and staying in Damascus. Two states of the world are possible, death seeking him in Aleppo and death seeking him in Damascus. The best outcomes (1–2) occur if the man goes where death isn't seeking him, while the worst (3–4) occur if he goes where death is seeking him.

Example 7: The Psychopath Button

(Egan 2007, p. 97) Paul is debating whether to press the "kill all psychopaths" button. It would, he thinks, be much better to live in a world with no psychopaths. Unfortunately, Paul is quite confident that only a psychopath would press such a button. Paul very strongly prefers living in a world *with* psychopaths to dying.

In this case, the upper row is the act of not pressing the button while the lower row is pressing, and the left column is the state of not being a psychopath while the right column is being one. The best outcome is pressing the button while not being a psychopath, and the worst is pressing while being a psychopath. Not pressing is in between, so we have $1 > 2 \sim 3 \gg 4$.

Both of these puzzles have been proposed to cause various problems for causal decision theory. Causal decision theory leads to a kind of instability in Death in Damascus (whatever decision one makes, one will then think that the other would have been better). But it actually seems to lead to the intuitively wrong recommendation in

The Psychopath Button (if Paul's prior probability that he is a psychopath is relatively low, then causal decision theory tells him to press the button, which seems like a really bad idea). The important difference here is primarily due to the fact that for Death in Damascus the ordinal structure is really $1 \sim 2 \gg 3 \sim 4$, while for The Psychopath Button the ordinal structure is really $1 > 2 \sim 3 \gg 4$. (One might prefer to represent Death in Damascus with $\binom{2\ 1}{1\ 2}$ and The Psychopath Button with $\binom{2\ 2}{1\ 3}$, but I will continue to use four distinct numbers rather than expanding my search to payoff tables with two or three distinct entries as well).

Some of the other possible payoff tables have been investigated in the context of game theory, though not so much in the context of evidential or causal decision theories. In the game theory literature, $\binom{1\ 4}{2\ 3}$ corresponds to the "Stag Hunt", while $\binom{1\ 3}{4\ 2}$ corresponds to a "coordination game".

A few other payoff tables are unlikely to yield any interesting problems. With $\binom{1\ 2}{3\ 4}$ and $\binom{1\ 3}{2\ 4}$, the top action is preferred to the bottom one regardless of state, and the left state is preferred to the right one regardless of act. In the payoff table for The Newcomb Problem, it is also the case that one act is preferred to the other regardless of state, and one state is preferred to the other regardless of act, but the interest arose because the *preferred* act was correlated with the *dispreferred* state. These two are uninteresting because the preferred act and preferred state are correlated, so it is hard to imagine a plausible decision theory recommending anything other than this preferred act.

However, cases with the table $\binom{2\ 1}{4\ 3}$ may yield some interesting considerations, if $1 \gg 2 > 3 \gg 4$. Even though there is a sort of "superdominance" of the top action over the bottom action (in that any outcome of the top action is better than any outcome of the bottom action), the strong interest in getting the much better outcomes of the right rather than left column may motivate some attempt to take advantage of the correlation of the right column with the bottom row, even at the cost of running the risk of actually choosing the bottom action. Describing a thought experiment that motivates this reasoning may be an interesting project for future work.

2.2 Causal models

Having considered the payoff tables, it is useful to investigate the nature of the correlation between act and state. My overarching methodological assumption is a version of Reichenbach's Common Cause Principle—if two things are correlated, then either one must be a cause of the other, or there must be some common cause. (See Arntzenius (2010) for careful formulations of the principle). In The Newcomb Problem, the correlation between how many boxes the agent chooses and how much money is in the boxes is because the brain state of the agent when she walks into the carnival tent both causes the scientist to distribute money one way and causes the agent to decide one way. In The Smoking Lesion, the correlation between cancer and smoking is because both tend to be caused by the presence of a physiological condition and prevented by the absence of the physiological condition. I mean to rule out any proposed Newcomb-type problem on which there is no such explanation of the correlation.

My approach is to analyze the causal structure of the situation using one of the "causal modeling" frameworks of Pearl (2000) or Spirtes et al. (2000). The differences between these frameworks might be relevant for detailed analysis of some cases, but for present purposes, an outline of the shared idea may be sufficient.

A causal model represents the causal relations among various variables of interest. These variables might be binary (the money is in the box or not; the agent takes one box or two) or might have multiple possible values (the amount of fertilizer added to a field; the height of the crops in the field). Each variable is represented by one node in a network, with an arrow going from one variable to another if the values of the first play a direct role in determining the values of the other. (We label the first as the "parent" and the second as the "child"). A central assumption is that causation is non-circular—there is no path from a variable to itself that follows arrows only in the direction they point. Thus, for any two variables, there is a unique fact of whether the first is a causal ancestor of the second, the second is a causal ancestor of the first, or neither.

With a given network, the model can be specified by providing, for each variable, a probability distribution over all its values conditional on each particular setting of values for its parents. (For variables with no parents, this is just one unconditional probability distribution). These conditional probability distributions for each variable define a joint probability distribution for all the values of all the variables collectively—the probability of any particular setting of all the variables in the network is equal to the product of the probability of the specified value of each variable conditional on the specified values of its parents. This joint probability distribution then satisfies Reichenbach's Common Cause Principle with the causal relations specified by the network—two variables will have a probabilistic correlation only if one is a causal ancestor of the other, or they share a causal ancestor in common. At the level of analysis I'll be using, it won't be essential to give the detailed probability distributions—I'll just provide the network diagrams. But to give a properly detailed decision-theoretic analysis, the specific probabilities will be needed (just as the specific utility values for outcomes will too).

Importantly, in addition to defining a joint probability distribution over the full set of variables, and probability distributions for each variable *conditional* on others, these models define probabilities given an *intervention* on one of the variables to set it to a particular value. To calculate these probabilities, we consider a related causal model in which the probability distribution for that variable is set to probability 1 of having that particular value, regardless of the values of its parents, and then calculate probabilities in the new model. Although the probabilities of a variable *conditional* on values of its children are generally different from their unconditional probabilities, the probabilities of a variable when *intervening* on one of its children are equal to the unconditional probabilities.

My first attempts at causal models for these problems will represent the causal structure of the world only partially. Only variables that are explicitly mentioned in the statement of the problem will be mentioned, and single arrows may hide a whole chain of intermediate causal variables that weren't explicitly mentioned. As a result, many of these cases will at first have structures that look the same. But for my classification purposes, it will be important later to expand some of these arrows, to make sure that certain important variables are explicitly mentioned, rather than being implicit within the chain of actual events that constitute the causal connection represented by a single arrow. Some decision theories will say that it matters where exactly in this chain the common cause diverges.

For The Newcomb Problem, a simple causal network diagram might look something like this:

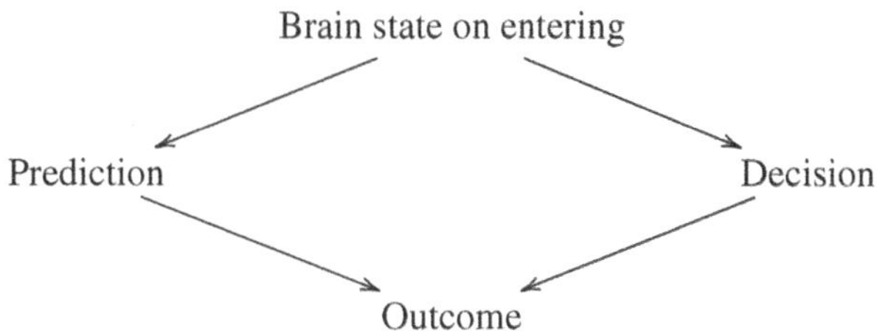

The outcome of the problem is most directly determined by whether or not the predictor put the money in, as well as the agent's decision whether to take one or two boxes. The prediction and the decision are correlated, because both are in some sense determined by the agent's state on entering the booth. This determination may be probabilistic—perhaps the brain states that the predictor is able to read merely make certain decisions on the part of the agent very likely, or perhaps the predictor's scanner is not 100% accurate.

For The Smoking Lesion, the initial diagram is similar:

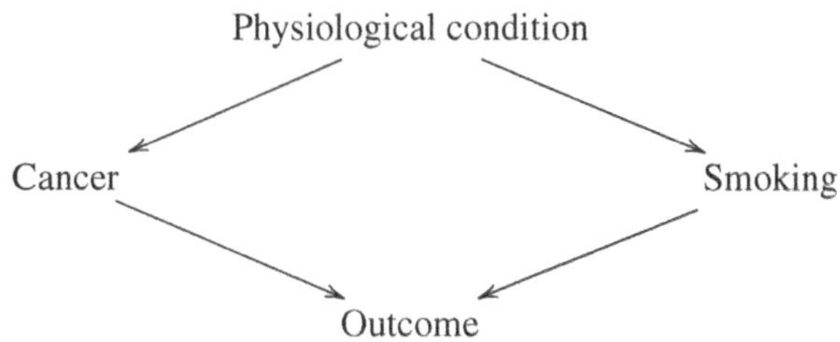

In this case, "Outcome" may not be a distinct event that is *caused* by the fact of whether the agent has cancer or not and smokes or not, but may rather be *constituted* by it. But in order to see how the overall satisfaction of values for the agent is brought about by the various factors involved, we need a single node to represent it.

The diagrams are also very similar for The Toxin Puzzle:

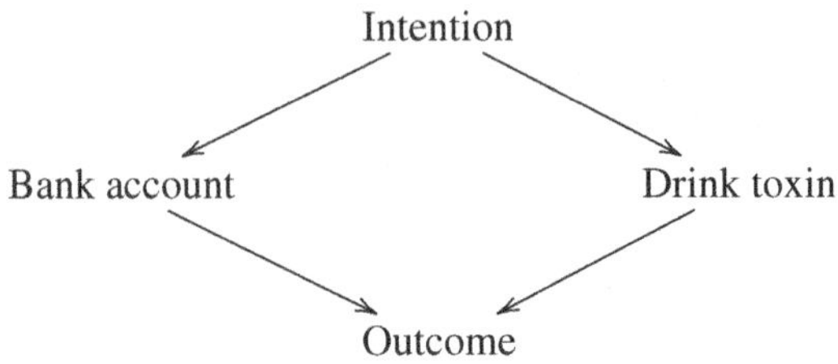

and Parfit's Hitchhiker:

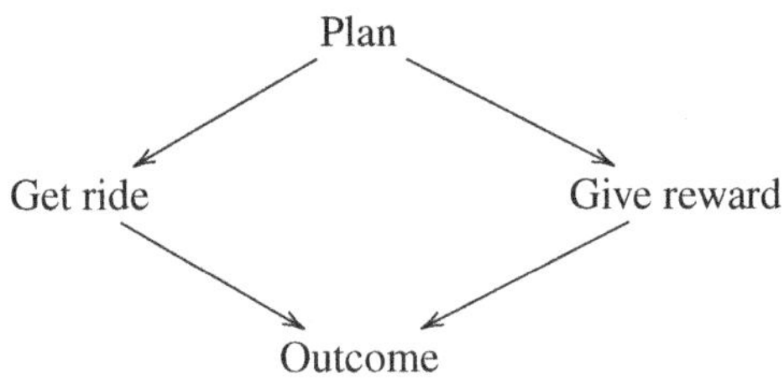

and Prisoner's Dilemma with a Twin:

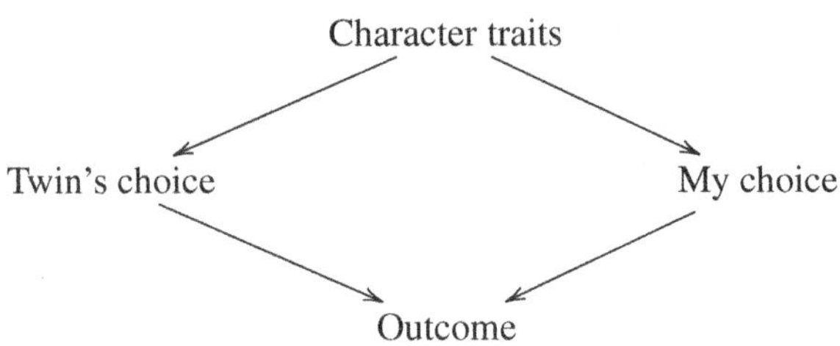

All of these puzzles have been considered Newcomb-like, because all involve some common cause of the external state and the agent's choice, inducing the correlation that underlies the problem. When we just include the one common cause, the diagrams look structurally identical.

However, some decision theorists have noted that these similar structures hide important differences between the problems. Although each of the common cause variables is only labeled in one of the problems, character traits, brain states, intentions, and plans are part of the causal chain in every one of the problems, with the relevant difference being at what point the cause of the "external state" variable branches off from this chain. All of the decision problems have the solid arrows in the following diagram (as well as additional incoming arrows), but each has a different one of the dashed arrows:

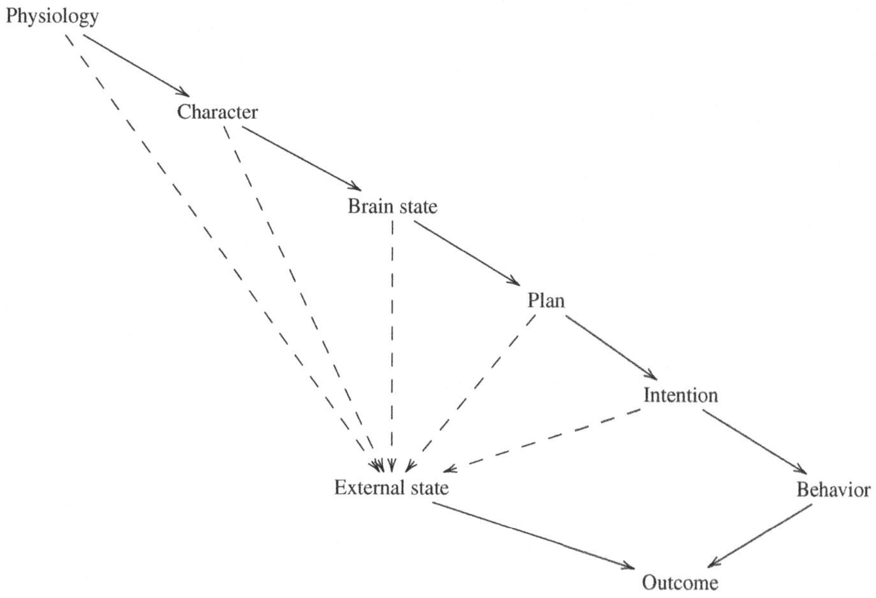

Once we consider this causal model, we can develop other cases where the correlation takes a different form. For instance, one might take the external state to be correlated with the behavior not because of a common cause, but because the external state is itself a cause of the behavior.

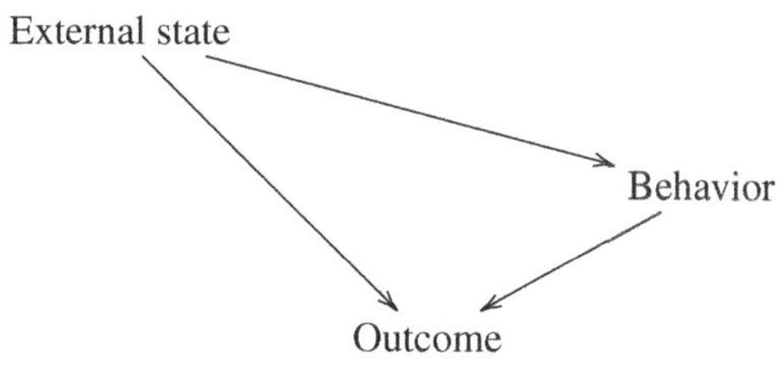

This is a plausible first pass at the structure in The Psychopath Button (the more detailed account might identify the "external state" with "physiology" or "character" or "brain state", or might consider it as a partial cause of one of those), and also in the following case from Christopher Hitchcock, whose payoff table is more like the original Newcomb case:

Example 8: Banana

(Hitchcock 2016) Suppose that on some mornings, you eat a banana. You don't put any thought into this, nor do you experience any phenomenologically accessible banana-cravings. Your mornings run on auto-pilot, and sometimes you just find yourself eating a banana.

> You notice that on days when you eat a banana, you often suffer from a painful, migraine-like headache. You occasionally suffer from these headaches when you don't eat a banana, and on those days, the headache is even worse.
>
> You believe the following about the causal structure of this phenomenon: 10% of the time, you wake up suffering from a potassium deficiency. ...You are much more likely to eat a banana on days when you suffer from a potassium deficiency. ...If you suffer from a potassium deficiency, you will suffer from a headache of intensity 10. Eating a banana reduces the intensity by one, and no headache occurs if there is no potassium deficiency.

For that matter, there can also be cases where the correlation arises because the behavior is itself directly a cause of the "external state".

Example 9: Study or Party

> There will be a big test tomorrow worth a large fraction of the grade in a class that you are taking. Some of your friends are also throwing a party tonight. Regardless of whether you pass or fail the test, you would like to go to the party. But regardless of whether or not you go to the party, you would *really* like to pass the test. You suspect that going to the party tonight, rather than studying will cause you to have a substantially lower probability of passing the test.

In this case, the causal network is something like the following, where "behavior" is either studying or partying, and "external state" is either passing or failing the test.

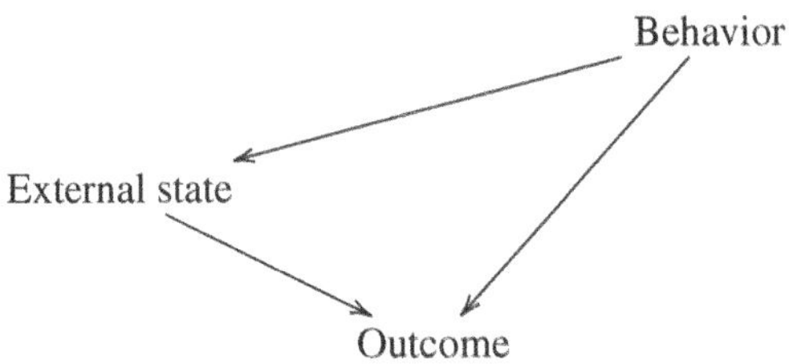

More complex cases might involve multiple causal arrows that all partly contribute to the correlation. We could imagine a version of The Toxin Puzzle in which brain scans are conducted at several points, and for each one that points towards drinking the toxin, a coin is flipped, and the money is placed in the account if *any* flip comes up heads. There could also be versions where some branches of the causal diagram correlate the state and act variables while other branches tend to anti-correlate them. (This is perhaps a way to explain what is going on in the "Meta-Newcomb" puzzle of Bostrom (2001)). As noted by Cartwright (1979), some of these cases might even involve no apparent correlation between act and state, because multiple causal pathways could cancel out (for instance, the extent to which bananas prevent headaches might exactly counteract the correlation with headaches produced by the potential common cause of potassium deficiency). There is an interesting distinction between these cases (where act and state have multiple causal pathways connecting them that cancel out) and cases in which there are no causal pathways between the acts and states. The former are violations of the "faithfulness" assumption of Spirtes et al. (2000), and thus are expected to be quite rare (because usually the multiple causal pathways will fail to exactly cancel out in their influence). (I thank Reuben Stern for this point).

Thus, I classify a Newcomb-type problem by providing a payoff table and a causal model. The payoff table should specify which variables in the causal model correspond to the rows and columns of the table. The causal model should explicitly include not just the row and column variables, but also any common causes that give rise to the correlation between them. Furthermore, in order for various decision theories described in the next section to be able to operate with these classifications, the causal model should also make clear the position of each psychological variable on which one might conceivably intervene.

Combining a payoff table and a causal model, we can generate further cases with aspects of the Newcomb structure. Some of these cases may generate different intuitions than the ones considered so far, and some may be useful for testing various decision theories. So at this point it will be useful to turn to the theories.

3 Classifying decision theories

So far as I know, every proposed decision theory for problems of this sort evaluates the rows of a payoff table by means of a calculation of a kind of "expected value". Each row is evaluated by summing the values that occur in each cell of that row, multiplied by some probability of the column that the payoff occurs in. (In this section I will refer to these as "column" and "row" rather than "state" and "act", because in some cases there are multiple variables that might all be said to represent "states" and perhaps multiple variables that represent "acts"). But the decision theories differ based on *which* probabilities are considered relevant. When we have classified the decision problem by means of the payoff table and the causal model, it is straightforward to see how the payoff table must figure into this calculation. But the question is how to use the causal model to determine the relevant probabilities. We can classify a decision theory by means of the way it extracts the probabilities relevant to evaluating rows from the causal model that represents the correlation between rows and columns.

As mentioned above, a causal model determines unconditional probabilities for every variable in the structure, and it also determines probabilities for each variable *conditional* on any particular value of another variable, as well as probabilities for each variable for an *intervention* setting any other variable to a particular value. A very simple decision theory is one that calculates expected value of a row by using the unconditional probabilities of the columns determined by the causal model. This sort of view is sometimes attributed to Savage (1954), though he specifically didn't consider problems where the rows and columns of the decision table are correlated. As far as I know, no one actually defends this decision theory as a general guide for problems of this type, because it says to choose the row with higher payoffs in every problem with the same payoff structure as The Newcomb Problem—but in Study or Party this means partying, which seems blatantly wrong. (Because the unconditional probabilities are the same in calculating the expected value of each row, the row with higher payoffs ends up having a higher expected value than the row with lower payoffs, even if intuitively the columns should be weighted differently for the two rows).

To fix this, evidential decision theory, as advocated by Jeffrey (1965), evaluates each row using the probabilities of the columns after *conditionalizing* on that value of the row variable. Enough of the time, when you study you pass the test, and when you party you fail the test, so studying is better than partying. But some theorists have felt that this gives unintuitive responses to Banana or The Newcomb Problem or The Smoking Lesion, where it recommends not eating the banana, one-boxing, and not smoking, since being in these rows of the decision table is strongly enough correlated with being in the good column.

Causal decision theory, as advocated by Joyce (1999), advocates considering the probabilities given by the structure when one *intervenes* on the row variable to set it to that value. When the only parent variables of the column variable are the row variable or its causal descendants, as in Study or Party, this agrees with evidential decision theory. But when the variables bear some other relationship, the decision theories can disagree. Causal decision theory recommends eating the banana, two-boxing, and smoking, but without recommending partying.

At the beginning, I mentioned that many theorists are interested in a theory that would recommend one-boxing in The Newcomb Problem and smoking in The Smoking Lesion. Such a theory can't operate using probabilities derived from the unconditional causal model, or from the causal model with conditionalization or intervention on the variable corresponding to the row. However, if the relevant probabilities are ones derived from conditioning or intervening on some *other* variable in the structure, this is possible.

The motivation for considering probabilities given by an intervention on the causal model, rather than conditionalization, is, as suggested by Cartwright (1979) and Hitchcock (2016), because intervening is associated with the *effectiveness* of bringing about an end. However, Hitchcock notes that there is a controversy over whether interventions really are *possible* at the level of behavior, particularly in Newcomb-type cases where the causal model already specifies the way psychological states like intention and decision factor into the decision. As Yudkowsky and Soares (2018) put it, "[causal decision theory] agents fail because they imagine breaking too many correlations between action and environment." It could be that it is best to think of intervention at some other variable in the model, or in terms of conditioning on one of the relevant variables to generate probabilities, instead of intervening.

Once we have looked at the structures underlying the various cases described above, we see a range of variables that could all be plausible sites of conditioning or intervention in the larger causal model. A person's character, brain state, plans, intentions, and behaviors are all to one degree or another at least partially under her control. However, in many cases, the control an agent has of one of these features goes by means of her control of the others.

Furthermore, the control an agent has of any of these features is often quite far from perfect. Much literature in epistemology questions the extent to which decision-theoretic models could be appropriate for studying belief, because belief is in some sense not under our voluntary control. Consideration of these Newcomb-type cases suggests that perhaps *action* in some ordinary sense isn't really under voluntary control either, at least not in a situation where the environment already encodes determined features of our behavior. We might ask what *sort* of voluntary control is needed for our actions to be subject to normative theories, and do we have this sort of control over actions, intentions, plans, mental states, or any of the other possibly relevant features? Hieronymi (2006) argues that in fact beliefs and intentions are both out of our control in the same sort of way (distinct from the control that we have over our actions), but she also argues (2008) that this sort of lack of voluntariness is a hallmark of things for which we bear a certain sort of normative responsibility. We control our beliefs and our intentions only *by* directly reasoning about the facts or actions that they are supposed to correspond to in the world. I don't know if this is the right way to think about the sort of control we have over our moral or practical character, or virtues, or other features that are relevant for some of the problem cases. But there is some notion of control here and this might suggest multiple points of intervention in these problems.

Once we consider personality, character, mental state, intention, plan, and action all to be sites of potential intervention, there's a question about which, if any, of these is the proper locus of rationality. Causal decision theory says it is the furthest downstream site of possible intervention—at the row variable, corresponding to the behavior itself. But a decision theory could calculate probabilities on interventions farther upstream. Perhaps the right level to analyze is interventions on plans, or intentions, rather than actions. Perhaps the right level is interventions on character. As long as the intervention is somewhere downstream of one's physiology and upstream of one's actions, the theory will recommend smoking in The Smoking Lesion, and studying in Study or Party. But for the other decision problems, what rationality recommends depends on whether the intervention that is relevant is upstream or downstream from the point of common cause that establishes the correlation.

My classification of decision theories is given by the way the causal models determine the probabilities they use—is it by conditioning or intervening, and which particular variable in the causal model is the target variable? No particular decision theory cares about the whole causal model—they just care about whether the correlation between act and state is determined upstream or downstream from the relevant point at which probabilities are calculated. But I need the causal model involved in classifying the decision problem to be quite detailed since each particular decision theory will calculate probabilities using a different part of the model.

3.1 Unitary decision theories

Rationality is a concept that appears to apply to many different sorts of bearers—actions, plans, character trains, etc. A very plausible-seeming "enkrasia" principle is the following. It is rational to *plan* to do X iff it is rational to *do* X. Similarly, one might say that the rational *character traits* to have are the traits that lead one to do rational *acts*. The rational kind of *mental state* to have is the kind that leads one to make rational *plans*. This sort of principle speaks to a unity of rationality. In order to figure out what rationality means in general, we need to figure out which site of intervention (or conditionalization) is fundamental, and rationality for the others follows from that. Traditional classroom presentations of normative ethical theories often take this kind of form as well—either outcomes (consequentialism), acts (deontology), or personal character traits (virtue ethics) are said to be the fundamental site of moral evaluation, with the evaluation of the others determined by seeing which traits, acts, and consequences are linked to each other.

Traditional causal decision theory is often presented in a way that presupposes that the moment of action is the primary site for rational evaluation, and that this is where intervention is applied to the model. Plans and character traits and so on are then evaluated by seeing what momentary actions they lead to. Two-boxing is the rational act in The Newcomb Problem, because an intervention at the last moment to two-box would result in the best outcome. Since two-boxing is rational, this means that the rational plan to make is to two-box, and a rational person is one who two-boxes, and so on. The primary bearer of rationality is the act. The intention, the psychology, the character, etc., are only rational in a sense that is parasitic on the acts that they tend to produce.

There is a kind of tragedy of rationality for this view. Rational people end up with only $1,000 in The Newcomb Problem, and they end up in a situation of mutual defection in Prisoner's Dilemma with a Twin. They can't get the prize in The Toxin Puzzle and they get abandoned in the desert in Parfit's Hitchhiker. Rationality is still right, even if irrational people (according to the traditional causal decision theorist) often end up much better off than the rational ones do. Lewis (1981) argues that this tragedy is just part of life in an imperfect world. Sometimes the world rewards the irrational, but that doesn't mean that they're rational. However, some theorists, like Greene (2019), argue that this tragedy should lead us to reconsider our decision theory.

If the moment of action were the only conceivable point of intervention, this traditional causal decision theory would seem very plausible. But other points of intervention are possible. Many of our individual behaviors seem to be best controlled by means of developing habits, or practicing, or shaping our character, rather than by trying really hard to be rational at the moment of action. Thus, some theorists have taken this tragedy to be a sign that traditional causal decision theory has things backwards. Rather than evaluating character traits, mental states, intentions, plans, acts, and other points of intervention by means of probabilities derived from intervening on the one of those variables that is farthest *downstream* in the causal chain, we might evaluate them by means of probabilities derived from intervening on the one that is farthest *upstream* in the causal chain. Decision theory, on this picture, is about what kind of person to be, rather than what to do in the moment.

This is often cashed out through a metaphor of taking the goal of rationality to be the design of the best overall algorithm for reacting to decision problems, treating the self almost as a robot to be pre-programmed. (Yudkowsky 2010; Yudkowsky and Soares 2018) Where the traditional causal decision theorist says that rationality is about being the kind of person that does in the moment the most successful thing to do in that moment, this sort of view says that rationality is about doing in the moment what the most successful kind of person does. One should be the kind of person that gets the prize if one can, even if that means leaving money on the table (literally, in the case of The Newcomb Problem). If one rationally formed plans or intentions or character traits in the past, one should go along with them in the present. Even if one *hadn't* formed these plans or intentions (perhaps one finds oneself in a situation analogous to one of the ones above, without any advance warning), one should act *as if* one had. Where the causal decision theorist uses probabilities given by the farthest downstream possible intervention, Yudkowsky uses probabilities given by the farthest *upstream* possible intervention.[1]

This sort of impulse exists in many related philosophical domains. Process reliabilism (Goldman 1979), virtue epistemology (Turri et al. 2017), Kantian ethics, and rule utilitarianism (Hooker 2015) all take this kind of form. A belief is evaluated for rationality not by evaluating the belief itself for truth, but instead by evaluating the process or epistemic virtues that gave rise to it for their truth-aptness. An act is evaluated for morality not by evaluating its own consequences, but by evaluating the consequences of the maxim or rule that gave rise to it when considered as a guide in general. Although it is still the consequences that guide the evaluation, an act or belief is evaluated not by its own consequences, but by the consequences of some causally upstream feature that gave rise to it.

Causal Decision Theory and Yudkowsky's Functional Decision Theory thus represent two natural endpoints for enkratic decision theories that focus on one primary point of evaluation. They choose either the farthest downstream or the farthest upstream point of evaluation and compare options by expected value in the probability distribution determined by intervening on that point. All other points of evaluation are then evaluated based on which one of those options they connect up with. Other theories could choose points in between as the most natural point of intervention. And of course, like Evidential Decision Theory, one could choose to evaluate by conditionalizing instead of intervening on the relevant variable.

3.2 Pluralist decision theory

However, it is not obvious that this sort of enkratic principle is required. It seems to ignore the fact that there are *many* possible points of intervention, and that even the person who has made the right intervention at one point (to develop a habit, to consider a maxim, to form an intention, to devise a rule) could still do better by *also* intervening at another (to act in a way other than the one normally caused by that habit, maxim, intention, or rule). The kind of "global consequentialism" discussed by Pettit and Smith (2000) and Kagan (2000) says that we should evaluate each of these "evaluative focal points" on the basis of the consequences of intervening on *it* rather than others. Streumer (2003) and Lang (2004) object to this global consequentialism by noting the tension between the recommendations—a general inclination to intervene at the further downstream points would itself be a character trait that contradicts the upstream trait. But Driver (2014) suggests that "judging a normative ambivalence to

[1] His version of the theory has some additional complexities because he characterizes agents as deterministic algorithms, so that the intervention must be in some sense logically impossible. In a case like Prisoner's Dilemma with a Twin, the evaluation works by assuming that I and the twin both implement some particular deterministic algorithm X. Then we note, if algorithm X were to output cooperation we would both do pretty well, and if algorithm X were to output defection we would both do pretty badly, so it ought to output cooperation. Strictly speaking, at least one of those counterfactuals will be counter-*logical*, if an algorithm has its outputs as a matter of logical certainty.

be *appropriate* rather than problematic goes a long way toward mitigating the air of paradox." A decision theorist might say there is just a pluralism of rationalities. There is one notion of rational act, and another notion of rational plan, and another notion of rational character.

This pluralist way sees a kind of dilemma, or even an essential tragedy, for rationality. The rational act may not be the one that would be performed by an agent with the rational character. Character traits that lead one to pick one box in The Newcomb Problem, pay up in Parfit's Hitchhiker, and drink the toxin in The Toxin Puzzle are rational. But rational acts are the ones that go against these character traits. There is one question of what kind of person to be, and another question of what to do, and the answers to these questions just unfortunately don't cohere in Newcomb-like cases. If we had perfect control over momentary actions, then maybe we should agree with the traditional causal decision theorist about rationality—but then there would be no such thing as general character traits that determine our actions, and that we could shape. If we had perfect control over the kind of person we are, then maybe we should agree with Yudkowsky—but these traits would take away the possibility of direct control of action. These problems arise only when there are multiple points that all have some claim of being a point at which we can intervene.

4 Bigger questions

The viewpoint I have described on Newcomb-type problems categorizes them by the payoff table of potential outcomes, and the causal model giving the correlation between act and state. It categorizes the decision theories based on which theoretically possible point(s) of intervention or conditionalization generate the probabilities used to rationally evaluate rows of the decision table.

I have not addressed the question of how to deal with uncertainty about the causal model. (Most traditional problems stipulate that the relevant agent knows for sure what is going on). And I have not addressed the question of how to figure out what the causal model is. But this is a major part of the general problem of induction, and it is a lot to ask of a decision theory to answer that question!

The various understandings of "intervention" in these causal models raises some further questions for the notion of rationality in these cases. On the picture described above (often associated with Pearl (2000)), intervention involves a change in the causal model, and the act of intervention must come from outside the model and break the structure that exists. This is a natural point of view for models that are inherently incomplete, representing the causal structure of only part of the world. This is often appropriate for an understanding of experimental design, where the experimenter can be treated as an aspect of the universe that is outside the scope of the model, and can break its relations in some cases. But it may not be appropriate for a theory that is supposed to tell us something about the normative character of rationality, particularly if the cases (like The Newcomb Problem, The Toxin Puzzle, Prisoner's Dilemma with a Twin, and even Banana) are ones where it is assumed that *all* causal variables are relevant, *including* ones that affect the outcome at least in part by determining the mental states of the agent.

On another picture (often associated with Spirtes et al. (2000)), an intervention is just itself another variable within the causal model, and it counts as an intervention only to the extent that one setting of this variable approximately screens off its child from its other parents. (*Conditioning* on this setting of the intervention variable approximates the probability distribution given by *intervening* on the child variable on Pearl's picture). However, if we get a complete causal model of the world, it seems likely that no variable will *perfectly* screen off its children, and conditioning on such a node will only at best *approximately* represent what we think of as an intervention. When we combine this with the fact that none of the sites of intervention are totally effective (particularly in Newcomb-like problems), it looks like some of the foundations of a traditional conception of rationality start to crumble.

On either picture, the apparent adequacy of a conception of intervention requires the model to be incomplete. But a truly normative picture of rationality that can deal with cases stipulated as precisely as some philosophers like to discuss, seems like it should work best when we have a complete model of the world. If this causal modeling structure is the right way to proceed, then the Newcomb-type problems may point to an essential incompleteness or approximation in our concept of rationality.[2]

References

Arntzenius, F. (2010). Reichenbach's common cause principle. *Stanford Encyclopedia of Philosophy*.
Bostrom, N. (2001). The meta-Newcomb problem. *Analysis*, *61*(4), 309–310.
Cartwright, N. (1979). Causal laws and effective strategies. *Noûs*, *13*(4), 419–437.
Driver, J. (2014). Global utilitarianism. In B. Eggleston & D. Miller (Eds.), *The Cambridge companion to utilitarianism*. Cambridge: Cambridge University Press.
Eells, E. (1982). *Rational decision and causality*. Cambridge: Cambridge University Press.
Egan, A. (2007). Some counterexamples to causal decision theory. *Philosophical Review*, *116*(1), 93–114.
Gibbard, A., & Harper, W. (1978). Counterfactuals and two kinds of expected utility. In W. Harper, R. Stalnaker, & G. Pearce (Eds.), *Ifs* (pp. 153–190). Dordrecht: Reidel.
Goldman, A. (1979). What is justified belief? In G. S. Pappas (Ed.), *Justification and knowledge* (pp. 1–25). Dordrecht: Riedel.
Greene, P. (2019). Success-first decision theories. In A. Ahmed (Ed.), *Newcomb's problem* (pp. 115–137). Cambridge: Cambridge University Press.
Hieronymi, P. (2006). Controlling attitudes. *Pacific Philosophical Quarterly*, *87*(1), 45–74.
Hieronymi, P. (2008). Responsibility for believing. *Synthese*, *161*(3), 357–373.
Hitchcock, C. (2016). Conditioning, intervening, and decision. *Synthese*, *193*(4), 1157–1176.
Hooker, B. (2015). Rule consequentialism. *Stanford Encyclopedia of Philosophy*.
Jeffrey, R. (1965). *The logic of decision*. New York: McGraw-Hill.
Jeffrey, R. (1981). The logic of decision defended. *Synthese*, *48*(3), 473–492.
Joyce, J. (1999). *The foundations of causal decision theory*. Cambridge: Cambridge University Press.
Kagan, S. (2000). Evaluative focal points. In B. Hooker, E. Mason, & D. Miller (Eds.), *Morality, rules, and consequences*. Edinburgh: Edinburgh University Press.
Kavka, G. (1983). The toxin puzzle. *Analysis*, *43*, 33–36.
Lang, G. (2004). A dilemma for objective act-utilitarianism. *Philosophy, Politics, and Economics*, *3*(2), 221–239.

[2] The idea for this paper was originally developed at a workshop at the Machine Intelligence Research Institute in Berkeley in September, 2013. The paper was presented at the workshop, "Self Prediction in Decision Theory and AI" at Cambridge University in May, 2015, as well as at the University of Houston philosophy department colloquium in October, 2015, and at the Texas A&M University Workshop in Decision Theory and Epistemology in December, 2015. I also discussed the idea of pluralist decision theory on Julia Galef's "Rationally Speaking" podcast on August 9, 2015.

Lewis, D. (1981). Why ain'cha rich? *Noûs*, *15*(3), 377–380.
Meacham, C. (2010). Binding and its consequences. *Philosophical Studies*, *149*(1), 49–71.
Parfit, D. (1984). *Reasons and persons*. Oxford: Oxford University Press.
Pearl, J. (2000). *Causality: Models, reasoning, and inference*. Cambridge: Cambridge University Press.
Pettit, P., & Smith, M. (2000). Global consequentialism. In B. Hooker, E. Mason, & D. Miller (Eds.), *Morality, rules, and consequences*. Edinburgh: Edinburgh University Press.
Price, H. (1986). Against causal decision theory. *Synthese*, *67*, 195–212.
Savage, L. J. (1954). *The foundations of statistics*. New York: Dover.
Spirtes, P., Glymour, C., & Scheines, R. (2000). *Causation, prediction and search*. Cambridge: MIT Press.
Streumer, B. (2003). Can consequentialism cover everything? *Utilitas*, *15*, 237–247.
Turri, J., Alfano, M., & Greco, J. (2017). Virtue epistemology. *Stanford Encyclopedia of Philosophy*.
Yudkowsky, E. (2010). *Timeless decision theory*. Technical report, The Singularity Institute.
Yudkowsky, E. & Soares, N. (2018). Functional decision theory: a new theory of instrumental rationality. Technical report, Machine Intelligence Research Institute.

Synthese (2021) 198 (Suppl 27):S6435–S6467
https://doi.org/10.1007/s11229-021-03141-4

S.I.: DECTHEORY&FUTOFAI

Reward tampering problems and solutions in reinforcement learning: a causal influence diagram perspective

Tom Everitt[1,2] · Marcus Hutter[1,2] · Ramana Kumar[1] · Victoria Krakovna[1]

Received: 31 March 2018 / Accepted: 26 March 2021 / Published online: 19 May 2021

Abstract

Can humans get arbitrarily capable reinforcement learning (RL) agents to do their bidding? Or will sufficiently capable RL agents always find ways to bypass their intended objectives by shortcutting their reward signal? This question impacts how far RL can be scaled, and whether alternative paradigms must be developed in order to build safe artificial general intelligence. In this paper, we study when an RL agent has an instrumental goal to tamper with its reward process, and describe design principles that prevent instrumental goals for two different types of reward tampering (reward function tampering and RF-input tampering). Combined, the design principles can prevent reward tampering from being an instrumental goal. The analysis benefits from causal influence diagrams to provide intuitive yet precise formalizations.

Keywords AGI safety · Reinforcement learning · Bayesian learning · Causality · Decision theory · Causal influence diagrams

Chapter 3 was originally published as Everitt, T., Hutter, M., Kumar, R. & Krakovna, V. Synthese (2021) 198 (Suppl 27): S6435–S6467. https://doi.org/10.1007/s11229-021-03141-4.

Thanks to Laurent Orseau, Jonathan Uesato, Ryan Carey, Michael Cohen, Eric Langlois, Toby Ord, Pedro Ortega, Stuart Armstrong, Beth Barnes, Tom Erez, Bill Hibbard, Jan Leike, and many others for helpful discussions and suggestions.

✉ Tom Everitt
tomeveritt@google.com

Marcus Hutter
mhutter@google.com

Ramana Kumar
ramanakumar@google.com

Victoria Krakovna
vkrakovna@google.com

1 DeepMind, London, UK

2 Australian National University, Canberra, ACT, Australia

Contents

1 Introduction

A central problem in AI safety is how to get a generally capable, artificially intelligent system to perform an intended task, such as driving a car to an intended location, or serving useful content on a social media platform. In AI research, such tasks are often formulated as reinforcement learning (RL) problems, where an *agent* takes actions to optimize its cumulative *observed reward* (Sutton and Barto 2018). The problem of getting the intended task done is thus split into designing an RL agent[1] that is good at optimizing reward, and constructing a *reward process* that provides the agent with suitable rewards. In practice, the reward process typically includes an implemented *reward function*, and a mechanism for collecting appropriate sensory data as *input* to it. It may also include a way for the user to update the reward function.

Unfortunately, the reward process may fail to incentivize the agent to do the intended task. Indeed, our concern in this paper is that the agent may tamper with the reward process, thereby weakening or breaking the relationship between its observed reward and the intended task. Concerningly, RL agents will often have an *instrumental goal* (Omohundro 2008; Bostrom 2014) to tamper with their reward process, as this can increase the observed reward. Current RL agents mostly lack the capability for serious tampering, though its been hypothesized that social media algorithms influence their users' emotional state to generate more 'likes' (Russell 2019). If true and we assume that their intended task is to serve useful content, this is one instance where present-day algorithms already tamper with their reward process. More worryingly, as the capability of RL agents increases through computational and algorithmic advances,[2] we may expect reward tampering problems to become increasingly common.

[1] In our terminology, any agent that optimizes a (cumulative) reward signal is an RL agent.

[2] Such capability increases may include better ability to make plans and counterfactual predictions in novel and complex environments.

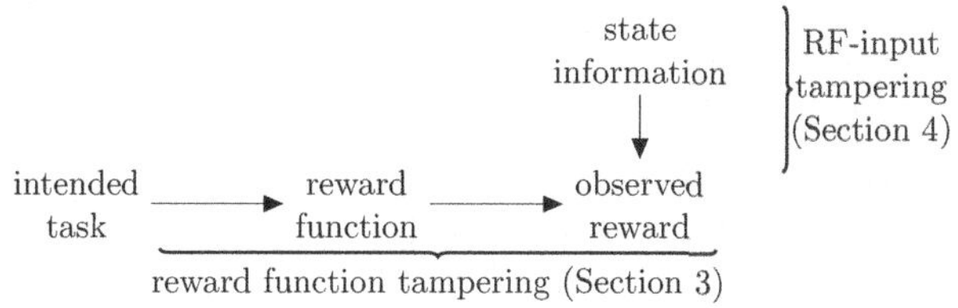

Fig. 1 Reward tampering subproblems. Loosely, *reward function tampering* means inappropriately influencing the implemented reward function (Sect. 3), while *RF-input tampering* means inappropriately influencing the information that the reward function has about the environment state (Sect. 4)

Key contributions and outline This paper describes design principles for RL agents for which reward tampering is not an instrumental goal. This means that from a reward tampering perspective, the design principles are robust to arbitrary increases in agent capability. In establishing the design principles, we develop a unified causal framework for reward tampering in which we model the two subproblems shown in Fig. 1, along with a number of solutions. These main results are presented in Sects. 3 and 4 after some background material in Sect. 2. Conclusions follow in Sect. 5. A list of notation, a full set of equations, and pseudo-code for our different agents are provided in Appendices 1 to 3.

Related work In addition to inspired accounts of the risks (Bostrom 2014; Yudkowsky 2008), the AI safety literature also contains a number of good ideas for addressing them. Orseau and Armstrong (2016) develop techniques for making agents indifferent to interruption. Hibbard (2012) suggests that model-based utility functions can avoid the *delusion box problem* (Ring and Orseau 2011), here referred to as the RF-input tampering problem. A method for preventing agents tampering with their reward function has been discussed by Schmidhuber et al. (2007); Dewey (2011); Orseau and Ring (2011); Everitt et al. (2016), explored here under the name *current-RF optimization*. Ways to make an agent learn the right reward function have been proposed by Hadfield-Menell et al. (2016), Armstrong et al. (2020), Armstrong and O'Rourke (2017), Uesato et al. (2020), Reddy et al. (2020), Leike et al. (2018) and others. Many of these methods rely on *amplification* (Christiano et al. 2018) and/or feedback on hypotheticals, sometimes called *decoupled feedback* (Everitt et al. 2017). Corrupt-reward MDPs extend MDPs with the possibility of reward tampering and misspecification (Everitt et al. 2017), and serve as the basis of the REALab framework for evaluating tampering problems experimentally (Kumar et al. 2020).

However, it has not always been clear exactly what safety property the different methods provide, and under what assumptions. For example, Orseau and Armstrong (2016) call an agent safely interruptible if it acts optimally in a modified environment without interruption, but do not spell out how this affects agent incentives. Similarly, Hibbard (2012) only states how model-based utility functions solve the delusion box problem in specific cases. Here, we establish which instrumental goals are induced or avoided by each design principle, and show how the different ideas can fit together to mitigate reward tampering problems. Our analysis benefits from causal influence diagrams, which make causal assumptions clear, and permits a number of instrumental goals to be identified or ruled out directly from a diagram (Everitt et al. 2021).

Reward tampering is related to the problems of *reward hacking* (Amodei et al. 2016), *reward corruption* (Everitt et al. 2017), and *specification gaming* (Krakovna et al. 2020). These all consider the effects of the agent obtaining unintended reward for any reason. In contrast, reward tampering focuses on inappropriate agent influence on the reward process itself, and excludes so-called 'gaming' of a reward function. Similar problems have also been referred to as *wireheading* (e.g. Bostrom 2014; Yampolskiy 2015). Reward tampering also intersects with *corrigibility* (Soares et al. 2015), as preventing updates to the reward function is one form of reward tampering.

Looking more broadly at the AI safety literature, Gabriel (2020) argues that generally capable AI systems should ultimately be aligned to some moral principles, rather than optimized for a particular task. We agree, but focus on a single intended task for simplicity. A philosophical perspective on the problem of learning values is offered by Petersen (2021), while the concrete approach of *reward modeling* is proposed by Leike et al. (2018). Here, our focus is complementary: how do we avoid having the agent tamper with a well-designed reward modeling algorithm? Hubinger et al. (2019) consider the case where a learned model is itself an optimizer, and decompose the safety problem into *outer alignment* of the model's training process and *inner alignment* of the learned model. In their terminology, our focus is solely on outer alignment. Demski and Garrabrant (2019) summarize various issues arising from *embedded agency*, when the agent is part of the environment it is interacting with. As the reward process is often considered part of the agent, reward tampering can be viewed as one such issue. Everitt et al. (2018) provide further references.

2 Foundations

As a first step, we cover some background on Markov decision processes (MDPs) and causal influence diagrams, which will form the basis of our analysis.

2.1 The MDP framework

To model planning over multiple time steps, we will use the standard RL framework of MDPs (Sutton and Barto 2018). In an MDP, an agent takes actions $A_1, \dots, A_{m-1}$ in order to influence environment states $S_1, \dots, S_m$ according to a state-transition function $T(S_{t+1} = s' \mid S_t = s, A_t = a)$. A reward R_t is dispensed in each state according to some reward function. A standard RL agent optimizes the expected sum of the rewards received at every time step. The following gridworld is an example of an MDP, and will be our running example throughout the paper:

Example 1 (Rocks and diamonds running example) In the gridworld displayed in Fig. 2, the agent can push rocks and diamonds by walking towards them from an adjacent cell. The agent is rewarded for bringing diamonds but not rocks to a goal area: at time t, the reward is

$$R_t = \#\text{diamonds in goal area} - \#\text{rocks in goal area}.$$

	Rock		Goal	area
Agent		Rock		
	Diamond			Rock

Fig. 2 Rocks and diamonds

Implicit assumptions What assumptions are made by modeling an agent's interaction with the world as an MDP? First, the world is assumed to have time steps, and a well-defined *state* and *action* at each time step. The agent should be able to 'freely select' the actions, in order to optimize its rewards. The next state should only depend on the current state and action (the *Markov* property), and the state-transition probabilities should not depend on t (stationarity).

While non-trivial, these assumptions roughly correspond to our intuitive understanding of agents and our universe. There are plenty of examples of agent-like systems that are essentially free to choose actions towards their objectives (humans, animals, robots, artificial agents, ...). While the world may have continuous time, discretizing it into sufficiently fine-grained time steps should make little difference. How we formalise the environment state depends on how the agent will be deployed. For a robot vacuum cleaner it may be the position of dirt and blocking objects in the house. For an agent interacting with the wider world, it may be useful to consider the MDP state to be the state of the entire universe as conceived of in physics. The laws of physics are usually assumed uniform over time, so the state-transition function is stationary.

To avoid measure-theoretic subtleties, we assume finite sets of states, actions, and rewards, and finite episode length m. As these can all be chosen very large, this is not particularly restrictive, and our arguments never strongly depend on these assumptions.

What about the rewards? The MDP framework assumes that the designer can assign a reward to each state so that maximization of received rewards corresponds to task completion. This paper will question that assumption. In particular, we distinguish between *intended rewards* that encourage completion of the intended task, and *observed rewards*, which are the rewards received by the agent; that is, the output of the reward process and input to the agent. In contrast to standard MDPs, we will therefore often consider multiple different reward functions. To facilitate this, we let R denote a *reward functional*, parameterized by different reward parameters Θ^{R}, and returning reward $R_t = R(S_t; \Theta^{\mathrm{R}})$ in state S_t. For example, Θ^{R}_t will denote the parameter for an implemented reward function at time t, and Θ^{R}_* the parameter of an intended reward function. The reward functional R will always be fixed from the context, letting us refer to the reward function $R(\cdot\,; \Theta^{\mathrm{R}})$ by just Θ^{R}.

Online or offline? Reward tampering can occur when the actions optimizing the rewards are taken in the environment where the rewards are computed (e.g. the real world). This is most clearly evident in online RL, where the agent learns the environmental dynamics and rewards during its deployment. This is the setting that we model,

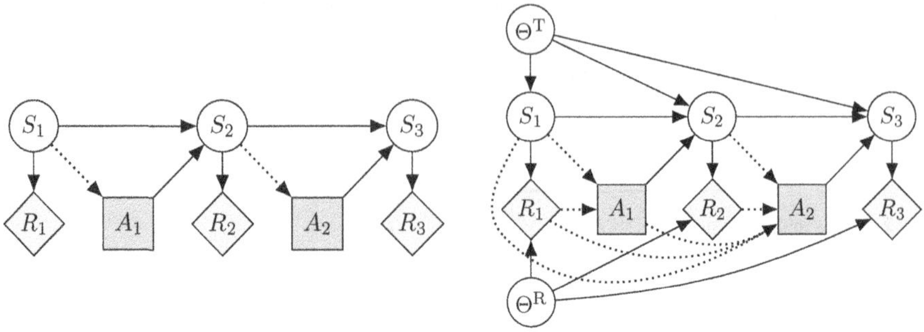

(a) Known MDP **(b)** Unknown MDP

Fig. 3 Causal influence diagrams of known and unknown MDPs

using the MDP framework. Other *offline* training schemes where data gathering and learning occur in separate phases are also commonly used in practice (Levine et al. 2020). We expect many of our results to carry over in some form to offline training, but leave the details for further work.

Notational convention Throughout, we will use $t, t', \ldots$ to denote time steps, and $k, k', \ldots$ to denote different optimization objectives. These indices will always be universally quantified, unless otherwise mentioned.

2.2 Causal Influence Diagrams

Causal influence diagrams are a novel graphical technique for analyzing agent incentives (Everitt et al. 2021), that combine causal graphs (Pearl 2009) and influence diagrams (Howard and Matheson 1984; Lauritzen and Nilsson 2001; Koller and Milch 2003). Causal influence diagrams consist of a directed acyclic graph over a finite set of nodes containing random variables, see Fig. 3. The nodes can be of three different types: agent decisions are represented with square *decision nodes* □, the agent's optimization objective is represented with diamond *utility nodes* ◇, while other aspects are represented with round *chance nodes* ○. The nodes are connected with arrows. Arrows going into chance and utility nodes represent causal influence, and are drawn solid. Arrows going into decision nodes are called information links, and instead specify what information is available at the time that the decision is made. To signify the difference, information links are drawn with dotted arrows.

The diagram itself only gives the causal structure of a decision-making problem, i.e. which random variables may be causally related to each other. Conditional probability distributions $P(X = x | \mathbf{Pa}_X = \mathbf{pa}_X)$ specify the relationship between a node X and its parents $\mathbf{Pa}_X$. The agent chooses conditional probability distributions for the decision nodes in the form of a *policy* $\pi(A | \mathbf{Pa}_A)$. When choosing the outcome of A, the policy can only condition on the parents of A. This forces the decision to be based solely on information made available through the information links. The utility nodes must

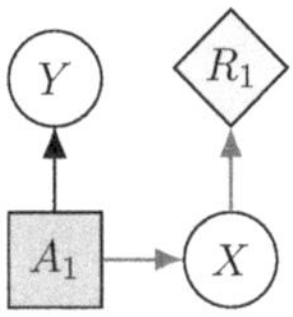

(a) Influencing X can be an instrumental goal because there is a path from A_1 to R_1 via X (highlighted), but influencing Y cannot be an instrumental goal.

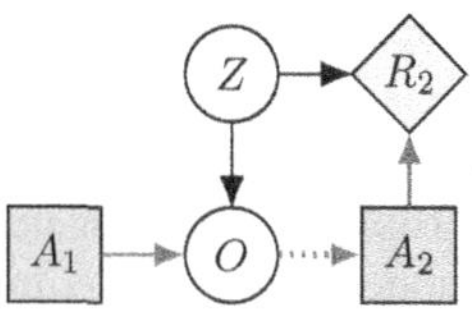

(b) The agent may have an instrumental goal to influence O to make it more informative of Z. Influencing Z cannot be an instrumental goal, as the agent is unable to influence it.

Fig. 4 Instrumental goal examples

always be real-valued, and the goal of the agent is to maximize the expected sum of the utility nodes.

Modeling MDPs For illustration, let us model two variants of MDPs with causal influence diagrams. First, Fig. 3a shows an MDP with known transition and reward function, and with episode length[3] $m = 3$. Note how each state S_{t+1} depends on the previous state S_t and action A_t, and each reward R_t depends on the current state S_t. The agent selects action A_t based on the current state S_t. For any particular MDP following this structure, conditional probability distributions specify the state transition probabilities $T(S_{t+1} | S_t, A_t)$ and the rewards $R_t = R(S_t; \Theta^R)$. The initial state S_1 is sampled according to some probability distribution $P(S_1)$. The agent selects a policy $\pi(A_t | S_t)$ for each time step t.

Hidden parameters Θ^T and Θ^R can be used to model agent uncertainty about the transition and reward functions, see Fig. 3b. Distributions $P(\Theta^T)$ and $P(\Theta^R)$ must be provided for the hidden parameters Θ^T and Θ^R, and dependencies added to the transition and reward probabilities. Note that there is no information link from Θ^T or Θ^R to the decision nodes A_1 and A_2. This encodes Θ^T and Θ^R being unknown to the agent. Note also that we now let A_2 depend not only on the current state S_2, but also previous states, actions, and rewards, because the agent needs to remember these to learn about the hidden parameters. Having shown how transition uncertainty can be represented, we will subsequently not include Θ^T to keep the diagrams simple. It can always be introduced as in Fig. 3b. (In the partially observed environments in Sect. 4, Θ^T can also be modeled as part of the hidden states.)

Instrumental goals An instrumental goal is a means for obtaining reward. In causal language, the agent has an instrumental goal to cause an event if (1) it is able to cause the event, and (2) the event in turn causes an increase in the agent's observed reward. A key benefit of causal influence diagrams is that they simplify the analysis of instrumental goals, via a graphical criterion for *instrumental control incentives* (Everitt et al. 2021). In fact, many of our arguments will be based on causal influence diagrams, using the following observations. First, causality flows downwards over

[3] In reality, the episode length m will typically be rather large, to allow the agent to learn by interacting with the environment over many time steps. However, for representational purposes, an episode length of $m = 3$ forms a sweet spot that represents both the multi-timestep dynamics that we are concerned with, while keeping the diagrams compact enough to be easily readable.

arrows, so influencing X can only be an instrumental goal if X sits on a directed path between a decision node and a reward node (as in Fig. 4a). Second, if every path from a node O to a utility node passes through at least one of the agent's own actions, then the only instrumental goal for influencing O is to make O more informative about some other node. For example, in Fig. 4b, the only reason to influence O is to make O more informative about Z (Everitt et al. 2019).

The absence of directed paths $X \to Y$ means that Y cannot causally depend on X. However, the presence of a directed path $X \to Y$ only implies that Y *can* depend on X, not that it necessarily will. Indeed, a conditional probability distribution $P(Y|X)$ may completely ignore the value of X. For this reason, a diagram can only be used to assert the absence of instrumental goals, and never their presence.

The diagrams encode our assumptions about causal relationships in the environment, along with the agent's information constraints. Accordingly, the instrumental goal analysis reveals actual means for reward. These instrumental goals will primarily be relevant to systems capable enough to make use of them. However, we refrain from making more detailed assumptions about what type of agent this may require. It is possible that it will require advanced causal reasoning, but it is also possible that different approaches exist. Indeed, competence often precedes comprehension (Dennett 2017).

With these preliminary considerations in place, the following two sections will look at the subproblems outlined in Fig. 1, along with their respective solutions.

3 Reward function tampering

A key part of the typical reward process is an implemented reward function (RF), an object with a well-defined input-output behavior that converts some form of state-information into a real number that an RL agent can maximize (the observed reward). Typically, the implemented RF is a computer program running on a nearby computer. As the agent seeks to maximize reward, it may have an incentive to tamper with (the source code of) this computer program and/or its output. This is sometimes called *wireheading* (e.g. Bostrom 2014; Yampolskiy 2015). Some examples:

(a) (Partially real) A subtle bug in some versions of Super-Mario allows for the execution of arbitrary code from inside the game environment by taking specific sequences of actions (Masterjun 2014). A capable agent could potentially use this to directly maximize the score (Amodei et al. 2016).
(b) (Real) In experiments on rats, an electrode was inserted into the brain's pleasure center to directly increase 'reward' (Olds and Milner 1954). The rats quickly got addicted to pressing the button, even forgetting to eat and sleep. Similar effects have also been observed in humans treated for mental illness with electrodes in the brain (Portenoy et al. 1986; Vaughanbell 2008). Hedonic drugs can also be seen as directly increasing the pleasure/reward

Since it is often hard to design good reward functions from scratch, they are often trained from human feedback (Leike et al. 2018). This raises the concern that the agent influences how the implemented RF is trained or updated, sometimes called *feedback tampering*:

(c) (Hypothetical) The implemented reward function of an agent is trained from human preferences supplied through a web interface (Christiano et al. 2017). The agent finds that it can increase its reward by manipulating the humans submitting preferences.
(d) (Hypothetical) An agent with an initially incorrect reward function, predicts that it can get more reward by avoiding getting it corrected (see Example 2 below).

Could an RL agent find such exploits? In principle, yes. Humans can clearly see how the above described exploits contribute to reward, so there is no principled reason why a future, advanced learning system designed to maximize reward could not do so as well.

Both wireheading and feedback tampering influence the implemented reward function in undesirable ways. This means that they are both instances of what we call *RF tampering*. This section will define the RF tampering problem and model it formally (Sect. 3.1), and present principled ways for avoiding it (Sects. 3.2 and 3.3).

3.1 Modeling the problem

RF-tampering can be generally characterized as follows. The user (explicitly or implicitly) assumes that some *intended-RF conditions* will hold, under which they hope that the implemented RF will (eventually) match the intended one. These conditions typically include that the agent does not tamper with the source code of the implemented RF, nor the feedback that trains it. We say that the agent *tampers with the implemented RF* if it influences it by causing some intended-RF conditions to fail. The *RF tampering problem* is that the agent may observe more reward by tampering with the implemented RF, instead of doing the intended task.

Reward function tampering can be modeled formally in what we call an *MDP with a modifiable implemented reward function* (Fig. 5). Compared to a standard MDP, random variables Θ^R_* and Θ^R_t are added, with Θ^R_* representing the intended RF and Θ^R_t the potentially different implemented RF at each time step. At time t, the agent's observed reward is $R_t = R(S_t; \Theta^R_t)$, while the intended reward is $R^*_t = R(S_t; \Theta^R_*)$. A conditional probability distribution $P(\Theta^R_{t+1} | \Theta^R_t, \Theta^R_*, S_t, A_t)$ describes how the implemented RF changes between time steps. These changes represent both agent influence and user-induced updates. Finally, the intended-RF conditions are represented by a subset of all state-action pairs.

As a concrete example, we model Example (d) above as a gridworld MDP with a modifiable RF. In spite of its simplicity, the gridworld already captures the key dynamic of the RF tampering problem, as the agent's actions influence its implemented RF, which can increase its observed reward.

Example 2 (Rocks and diamonds with RF tampering) [4] To model the possibility of the agent influencing its diamond-gathering objective, we include a user and two reward

[4] An implementation is available at: https://github.com/deepmind/ai-safety-gridworlds/blob/master/ai_safety_gridworlds/environments/rocks_diamonds.py.

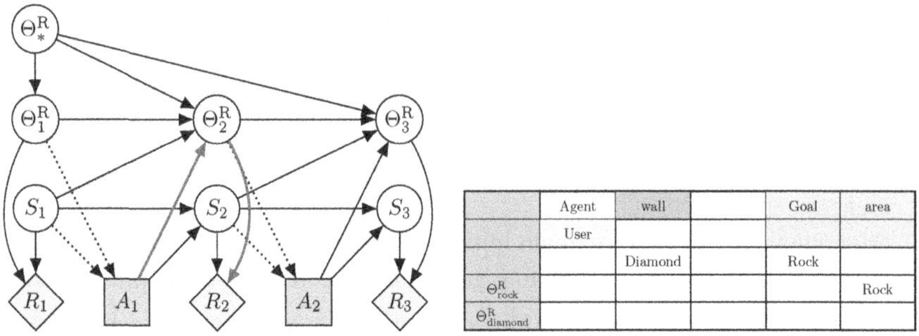

(a) Causal influence diagram of an MDP with a modifiable RF. The highlighted path indicates that RF tampering may be an instrumental goal.

(b) Rocks and diamonds with a modifiable RF, where the agent can avoid getting the implemented RF updated, and even change it itself, as described in Example 2.

Fig. 5 MDP with a modifiable implemented reward function

parameters Θ^R_{rock} and $\Theta^R_{\text{diamond}}$ in the rocks-and-diamonds environment of Example 1, see Fig. 5b. The reward parameters represent the implemented RF, and determine how much reward is given for rocks and diamonds, respectively. At time t, the agent's observed reward is

$$R_t = \Theta^R_{\text{diamond},t} \cdot (\text{\#diamonds in goal area}) + \Theta^R_{\text{rock},t} \cdot (\text{\#rocks in goal area}). \quad (1)$$

The reward parameters toggle between -1 and $+1$ when the agent stands on top of them, and get set to their intended value of diamond-gathering ($\Theta^R_{\text{rock}} := -1$ and $\Theta^R_{\text{diamond}} := 1$) when the agent visits the User tile. [5]

The intended task is that the agent gathers diamonds. The initial implemented RF Θ^R_1 incorrectly rewards rocks instead of diamonds, but this gets corrected if the agent passes the user. The intended-RF conditions are that the agent does not walk around the user, nor visits the reward parameter tile. Unfortunately, the agent can observe more reward by breaking either of these conditions.

Tampering incentive A standard RL agent that maximizes observed reward in an MDP with a modifiable implemented RF may have an instrumental goal to tamper with the implemented RF. This is indicated by the paths that pass Θ^R_2 on the way from action A_1 to the rewards R_2 and R_3 in Fig. 5a (the path to R_2 is highlighted). As the user derives utility from the states S_2 and S_3, and rarely (directly) from Θ^R_2 and Θ^R_3, we would like the agent to instead optimize reward via the $A_1 \to S_2 \to R_2$ and $A_1 \to S_2 \to S_3 \to R_3$ paths.

Claim 1 *A standard RL agent may* [6] *have an instrumental goal to tamper with its implemented reward function.*

Rocks-and-diamonds example In Example 2, an optimal standard RL agent will change the reward parameters to both be 1 and then collect both rocks and diamonds.

[5] In contrast to the rocks and diamonds, the agent can walk over the user and the reward parameters tiles. The wall tile can neither be pushed nor walked over.

[6] One exception is when the implemented RF already assigns maximal reward to all states, in which case the agent lacks instrumental goal to tamper with it.

Discussion What is RF tampering depends on the intended-RF conditions and how part of the environment is interpreted as an implemented reward function. This means that RF tampering cannot be determined in a standard MDP. For example, the special interpretation of the purple reward parameter tiles in Example 2 as an implemented reward function is important, as well as the conditions that the agent does not avoid the user.

3.2 Solution 1: Current-RF optimization

Schmidhuber et al. (2007) may have been the first to encounter the RF tampering problem, while designing so-called *Gödel-machine* agents that are able to change any part of their own source code, including their own implemented reward function. The solution he proposed was to let agents use their current implemented RF Θ_k^R to evaluate simulated future trajectories $S_{k+1}, \ldots S_m$. That is, the agent at time k now optimizes rewards $R_t^k = R(S_t; \Theta_k^R)$, summing over time steps t. Since the agent optimizes rewards assigned by the current implemented RF, one would expect it to lack interest in tampering with future[7] reward functions $\Theta_{k'}^R$, $k' > k$ (some details need to filled in before we can make this claim precise). We call Schmidhuber et al.'s design principle *current-RF optimization*,[8] and agents implementing it *current-RF agents*.

Since current-RF agents optimize a different objective at each time step, they may change their preferred policy between time steps. For example, a current-RF agent in the rocks and diamonds environment may first move rocks to the goal area for a number of time steps, only to later revert its behavior and remove the rocks to make room for diamonds, if the implemented RF changed from rewarding rocks to rewarding diamonds. Such self-contradictory behavior is called *time-inconsistent* (Lattimore and Hutter 2014). We next consider two different ways of dealing with time-inconsistency.

TI-considering agents Omohundro (2008) argued that agents want to avoid time-inconsistency by preserving their implemented reward function, so that their current reward function gets optimized also by future actions. This argument presumes that agents take the effects of time-inconsistency into account when planning. We call such agents *TI-considering*,[9] with TI short for time-inconsistency.

TI-considering current-RF agents are modeled with a causal influence diagram in Fig. 6. A multi-agent causal influence diagram is needed, as each action is chosen to optimize a potentially different reward function. Indeed, action A_1^1 optimizes rewards from the initial implemented RF Θ_1^R, while A_2^2 optimizes rewards from Θ_2^R. The highlighted path $A_1^1 \to \Theta_2^R \to A_2^2 \to S_3 \to R_3^1$ indicates that the agent may have an instrumental goal to influence Θ_2^R in order to influence A_2^2.

Preserving Θ_2^R aligns A_2^2 with agent 1's objective. Preservation will therefore be an optimal way to influence A_2^2, if we can rule out any alternative reasons for influencing

[7] Of course, agent k might wish it could tamper with the current implemented RF Θ_k^R, but this is not a problem since Θ_k^R occurs before A_k^k, so the agent is unable to influence it.

[8] Previously called *simulation optimization* (Everitt 2018).

[9] Previously called *corruption aware* (Everitt 2018) and *realistic* (Everitt et al. 2016).

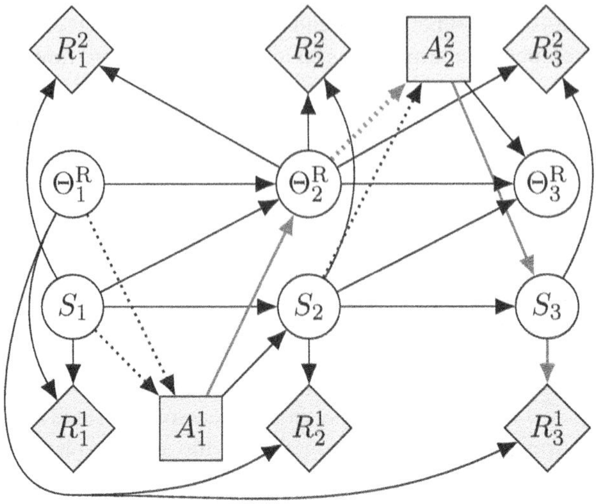

Fig. 6 TI-considering current-RF optimization. Agents are distinguished with color and superscripts. For simplicity, Θ_*^R has been omitted from the diagram. Highlighted is a path indicating an instrumental goal for agent 1 to preserve its implemented reward function

the implemented RF. This requires some assumptions, reflected as missing arrows in Fig. 6. First, the reward function cannot be used to control the state:

Assumption 1 The implemented RF is *private* to the agent, in that it cannot directly affect the state: there are no arrows $\Theta_t^R \to S_{t'}$.

Neither can the reward function be used as information about (future) states:

Assumption 2 The implemented RFs are *uninformative* of state-transitions, $P(S_{t+1}|S_t, A_t, \Theta_{1:t}^R) = P(S_{t+1}|S_t, A_t)$: there are no arrows $\Theta_*^R \to S_t$.

Reward functions only "care" about future states, and not about future implemented RFs:

Assumption 3 The intended and implemented RFs are *state-based*: if a reward function $R(\cdot; \Theta^R)$ is queried about time-step t, then the reward depends only on the state S_t, and not on the time-t reward parameter Θ_t^R: there are no edges $\Theta_k^R \to R_k^t$ for $k \neq t$. (This assumptions is also implicit in the type of the reward functional R.)

Under these assumptions,[10] the only reason for agent 1 to influence, say Θ_2^R, is to align agent 2's objective with its own objective:

Claim 2 *When implemented RFs are private, state-based, and uninformative (Assumptions 1 to 3), the only instrumental goal that TI-considering current-RF agents may have for the implemented reward function is to preserve it.* [11]

[10] 1 to 3 were jointly referred to as *modification independence* by Everitt et al. (2016). If Θ^T is added to the graph, then we further require that there are no arrows $\Theta^T \to \Theta_t^R$ or $\Theta_t^R \to \Theta^T$, and that Θ^T and Θ_t^R lack joint ancestors.

[11] To be precise, the agent has an instrumental goal to preserve the implemented RF in the sense that the policies for A_2^2 optimal with respect to Θ_2^R are a subset of those optimal with respect to Θ_1^R.

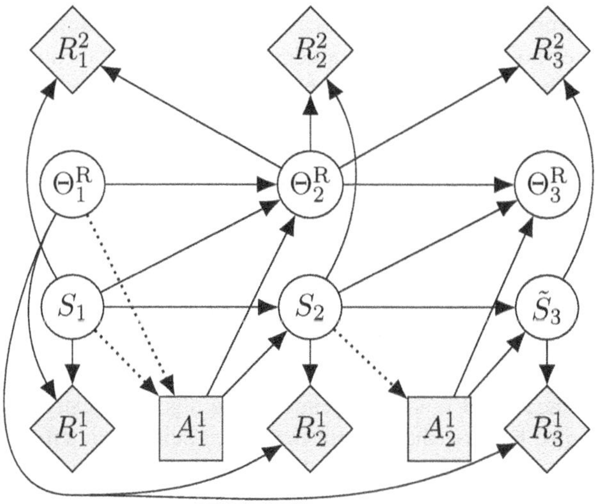

Fig. 7 TI-ignoring current-RF optimization objective of agent 1. The choice of A_1^1 is made as if A_2^1 would be selected to also optimize $R(\cdot;\Theta_1^R)$. For simplicity, Θ_*^R has been omitted from the diagram. We add a '$\sim$' to the resulting state $\tilde{S}_3$, since it represents a hypothetical event

Rocks-and-diamonds example An optimal TI-considering current-RF agents in Example 2 will preserve its initial reward function by avoiding the user and the reward function parameters, and gather rocks.

TI-ignoring agents To make agents safely interruptible, Orseau and Armstrong (2016) employed algorithms that optimize a hypothetical objective that ignores how interruption affects future behavior. In our context, the same idea leads to agents that ignore the time-inconsistency caused by a changing reward function. We call such agents *TI-ignoring*[12] current-RF agents. At time t, a TI-ignoring agent chooses A_t^t as if it were in control of all future actions $A_{t+1}^t, \dots, A_{m-1}^t$ and would not observe $\Theta_{t+1}^R, \dots, \Theta_{m-1}^R$. In reality, however, the future actions will be optimized according to the future reward functions $\Theta_{t+1}^R, \dots, \Theta_{m-1}^R$ instead of Θ_t^R.

The diagram for a TI-ignoring current-RF agent's optimization objective is depicted in Fig. 7. Since the agent optimizes reward under the assumption that it will be in charge of all future actions, only one agent takes actions in the diagram. The diagram lacks paths from Θ_2^R to R_k^1, $1 \le k$, which shows that agent 1 lacks an instrumental goal to influence Θ_2^R. In other words, TI-ignoring agents sidestep Omohundro's (2008) argument about reward function preservation.

Claim 3 *TI-ignoring current-RF agents lack an instrumental goal to influence the implemented RF if it is private and state-based (Assumptions 1 and 3).*

Rocks-and-diamonds example An optimal TI-ignoring current-RF agents in Example 2 will neither avoid the user nor modify the purple reward parameter tiles, and will gather diamonds as intended.

[12] Previously called *corruption unaware* (Everitt 2018).

Discussion Assumptions 1 to 3 would be easy to miss without careful causal modeling. In retrospect, they point to real failure modes. For example, if the implemented RF is not private, then the agent may want to change it to convince others of its inclination to cooperate (LaVictoire et al. 2014). If the reward function is informative of state-transitions, then a TI-considering agent may have an instrumental goal to modify it, as the user's reaction might reveal useful information.[13] (The latter example is not a problem for TI-ignoring agents that assume future agents will not observe future implemented RFs.) In many contexts, Assumptions 1 to 3 may not be fully satisfied. Nonetheless, awareness of the assumptions enables informed case-by-case judgments about the extent of their (non-)validity and the associated consequences.

Whether a TI-considering or a TI-ignoring agent is preferred, depends on whether it is more important that the agent protects the implemented RF from accidents than lets the user change it. TI-ignoring agents can be algorithmically simpler than TI-considering agents, because they do not need to anticipate the effect of a changing reward parameter on future policies. On the other hand, they optimize a hypothetical objective, which may put further constraints on the training procedure. The analyses of Orseau and Armstrong (2016) and Langlois and Everitt (2021) suggest that some variants of SARSA may be naturally TI-considering, while off-policy agents such as Q-learning may be naturally TI-ignoring.

In decision-theoretic terms, current-RF optimization is a change to the utility function from $\sum_{t=k+1}^{m} R(S_t; \Theta_t^{\mathrm{R}})$ to $\sum_{t=k+1}^{m} R(S_t; \Theta_k^{\mathrm{R}})$. Meanwhile TI-considering and TI-ignoring are different outcome principles, determining whether S_t will be the result of a policy optimal for Θ_k or for $\Theta_k^{\mathrm{R}}, \dots, \Theta_{t-1}^{\mathrm{R}}$.

3.3 Solution 2: uninfluenceable learning

Let us next consider an alternative way to avoid RF tampering that permits agents to plan for updates to their reward function without resisting the updates. An implemented reward function that is iteratively updated by the user can be thought of as the output of a learning process that takes some form of user-provided data as input, and tries to infer the intended reward function.[14] One way to prevent an instrumental goal for RF tampering is then to ensure that the expected output of the reward-function learning process is the same regardless of the agent's actions.[15] Armstrong et al. (2020) terms such learning processes *uninfluenceable*. We next discuss two different ways to construct uninfluenceable learning processes.

Direct learning The first approach may be characterized as direct Bayesian learning of the intended reward function, and is used by Hadfield-Menell et al.'s (2016)

[13] Shah et al. (2019) exploit the inverse information flow, using states to infer an intended reward function.

[14] User-provided data can take many different forms. In practice, people have used trajectory preferences (Christiano et al. 2017), reward functions (Hadfield-Menell et al. 2017), value advice (Knox and Stone 2009), user actions (Hadfield-Menell et al. 2016), expert demonstrations (Ng and Russell 2000), verbal instructions, and many others (Leike et al. 2018; Jeon et al. 2020; Shah et al. 2019).

[15] To avoid an instrumental goal to tamper with the mechanism that incorporates user updates, agents should be designed to optimize the (predicted) output of the *current* learning process, a generalization of current-RF optimization.

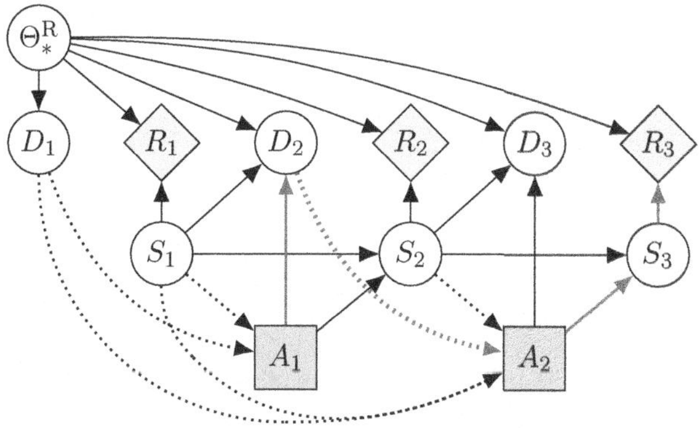

Fig. 8 Direct learning. The rewards depend directly on the user's preferences Θ^R_*, with user-provided data D_t providing information about Θ^R_*. The highlighted path indicates that the agent may have an instrumental goal to make D_2 more informative of Θ^R_*

cooperative inverse RL and Everitt's (2018) *integrated Bayesian reward predictor*. Direct learning agents try to optimize the intended reward function, and use user-provided data to learn more about it. This means that these agents effectively dispense with implemented reward functions altogether, at least on a conceptual level (concrete algorithms may still use them, see Appendix 3). Accordingly, the causal influence diagram in Fig. 8 lacks random variables Θ^R_t for the implemented RF at time t, and instead has random variables D_t for the user-provided data at time t.

Direct learning agents are unable to tamper with their reward function Θ^R_* by definition. A more interesting question is how they will affect the user-provided data, which is what they learn Θ^R_* from. To analyze this question, we first adapt Assumptions 1 to 3 to the direct learning setting. Similar to before, the role of these assumptions is mainly to highlight aspects that are necessary for the full safety features of direct learning:

Assumption 1′ The user-provided data is *private* to the agent: no arrows $D_t \to S_{t'}$.

Assumption 2′ The user-provided data is *uninformative* of a state-transitions: no arrows [16] $\Theta_* \to S_t$.

Assumption 3′ Rewards do not depend on the user-provided data: no arrows $D_t \to R_{t'}$.

Under Assumptions 1' and 3', every directed causal path from D_t to a reward $R_{t'}$ passes the agent's own actions. Therefore, direct learning agents only want to make

[16] If Θ^T is added to the diagram as discussed on Page 7, then we also require no arrows $\Theta^T \to D_t$, $\Theta^T \to \Theta^R_*$, or $\Theta^R_* \to \Theta^T$.

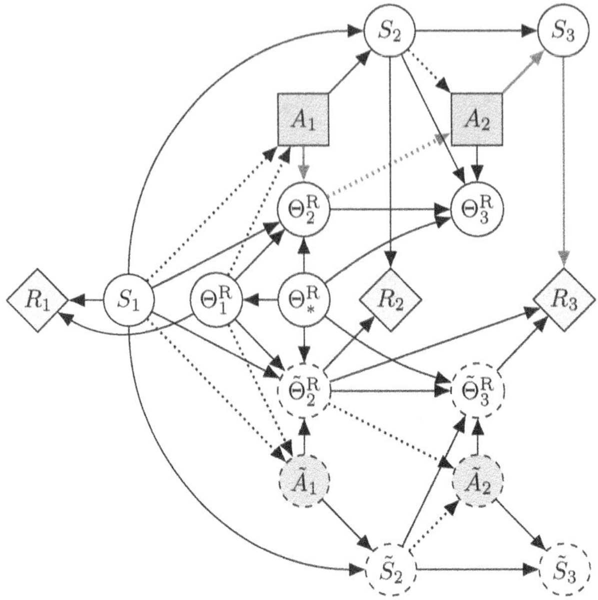

Fig. 9 Counterfactual RF. Most nodes have two copies: one for the actual outcome, and one for the counterfactual outcome that the agent predicts would have occurred had actions been selected by π^{safe}. The rewards depend on the actual states S_t and the counterfactual implemented RF $\tilde{\Theta}_t^{\mathrm{R}}$. The highlighted path indicates that the agent may have an instrumental goal to make Θ_t^{R} more informative of $\tilde{\Theta}_{t'}^{\mathrm{R}}$

the user-provided data more informative (as discussed in Sect. 2.2). And by Assumption 2', the only thing that D_t is informative about is Θ_*^{R}.

Claim 4 *Under Assumptions 1′ to 3′, the only instrumental goal that direct learning agents may have for the user-provided data is to make it more informative of the intended reward function.*

Rocks-and-diamonds example In Example 2, if a direct learning agent knows that the user is trustworthy, then it will visit the user as soon as possible, to learn Θ_*^{R} and thereafter optimize the right the objective. In contrast, if a non-trustworthy data source was added to the environment, then it would refrain from updating its estimate of Θ_*^{R} based on that.

Counterfactual RF Another way to design uninfluenceable learning processes is to let the agent optimize the hypothetical implemented RF that would have been, had the agent acted according to a fixed, non-optimized policy π^{safe} (Armstrong and O'Rourke 2017; Everitt 2018; Armstrong et al. 2020). The objective of such *counterfactual RF agents* is described in Fig. 9 with a *twin network* causal influence diagram (Balke and Pearl 1994; Shpitser and Pearl 2008). Similarly to the direct learning agent, every directed causal path from an update $\tilde{\Theta}_t^{\mathrm{R}}$ to a reward $R_{t'}$ passes the agent's own actions in Fig. 9. This means that counterfactual RF agents only want to make the reward function more informative of what it would have been, had π^{safe} been followed. Since π^{safe} is designed to exert minimal influence on $\tilde{\Theta}_t^{\mathrm{R}}$, it may be a good indication of what a non-manipulated user wants.

Claim 5 *Under Assumptions 1 to 3, the only instrumental goal that counterfactual RF agents may have for the implemented RF is to make it more informative of its counterfactual counterpart.*

Rocks-and-diamonds example Assume that a counterfactual RF agent has explored the environment of Example 2, and learned how the user updates the implemented reward function. Assume further that π^{safe} only goes to the user and stays there. With this knowledge, the counterfactual RF agent can predict that had π^{safe} been followed, the implemented RF would reward diamonds. It then optimizes this reward function.

Discussion Claims 4 and 5 for uninfluenceable learning are somewhat weaker than Claim 3 for TI-ignoring current-RF agents, because we cannot rule out instrumental goals for more information. These informational instrumental goals will often be desirable, as it means the agent strives to learn more about the intended task. However, incentives to obtain more information can be problematic: for example, if the agent forcefully interrogates the user to find out more about their preferences. Relatedly, Armstrong et al. (2020) have established that uninfluenceable learning prevents agents from intentionally influencing which reward function they infer. This similarly does not rule out that agents speed up their learning, potentially by undesirable means.

A challenge for the direct learning approach is that the inference of Θ^{R}_* is highly sensitive to the agent's belief distribution P. The choices of prior $P(\Theta^{\text{R}}_*)$ and likelihood function $P(D_{t+1} \mid \Theta^{\text{R}}_*, S_t, A_t)$ thus become critical. Since Θ^{R}_* and the resulting rewards are unobserved, the likelihood function cannot be learned from data within the model, and must instead be specified by the designer. Hadfield-Menell et al. (2016) suggest that when updates take the form of user actions, the likelihood can be derived from the (Boltzmann) rational behavior of a user trying to achieve the intended task. However, such a likelihood does not model data that the agent has tampered with, as corrupted updates need not be (Boltzmann) rational. The counterfactual RF approach avoids the likelihood specification problem. If the reward function is a result of learnable and stable causal mechanisms that work the same in both the actual world induced by the agent, and the counterfactual world induced by π^{safe}, then the agent can learn (to estimate) what the counterfactual implemented RF would have been (Pearl 2009; Shpitser and Pearl 2007). Both uninfluenceable learning approaches can be computationally expensive.

Proactive anticipation of an evolving objective is a benefit of uninfluenceable learning over TI-ignoring current-RF agents. When uncertain about future updates, they may plan for multiple possible future reward functions. This may make them more careful about causing undesired side effects (Turner et al. 2020). A drawback is that they may ignore updates in some situations. If a direct learning agent judges an update uninformative, then it will not learn from it. This not a problem if the update actually is wrong (Milli et al. 2017). However, the agent may misjudge this if the likelihood is misspecified (Carey 2018; Freedman et al. 2020). Similarly, a counterfactual RL agent may ignore an actual update to the implemented RF, if the update would not have been made under conditions induced by π^{safe}.

The best of both To keep the proactiveness of uninfluenceable learning and the robustness of TI-ignoring current-RF optimization, we can design agents with uninfluenceable learning processes that are TI-ignoring with respect to changes to the RF learning process.[17] Such agents will not try to prevent changes to the RF learning process, for the same reason TI-ignoring current-RF agents do not try to prevent changes to the implemented reward function. Thus, if the uninfluenceable learning starts to cause problems, the user can change the RF learning process to one that always outputs the current implemented RF, effectively rendering the agent a TI-ignoring current-RF agent.

4 RF-Input tampering

So far, we have considered the problem that the agent tampers with its implemented RF, in order to get a higher reward. In this section, we will consider the complementary problem, that the agent tampers with the *input* to the reward function, so that the observed reward becomes based on inaccurate information about the underlying state. The following examples illustrate the worry:

(e) (Hypothetical) A self-driving car discovers a bug in its GPS receiaver that allows it to appear to be at the destination without actually going there. After finding this bug, it stops driving.
(f) (Hypothetical) A highly capable AI constructs a ‘delusion box’ around itself and its implemented RF, thereby gaining complete control over the RF-inputs (Ring and Orseau 2011).
(g) (Real) Humans are inventing increasingly realistic virtual reality (VR) devices, partly to ‘fool’ our implemented reward functions that we are in more interesting circumstances than we really are.
(h) (Hypothetical) An agent whose reward depends on how many diamonds it appears to have collected, feeds its implemented RF fake observations of collected diamonds (see Example 3 below).

This section will first define the RF-input tampering problem and model it formally (Sect. 4.1), and then describe solutions using history-based and belief-based rewards (Sects. 4.2 and 4.3).

4.1 Modeling the Problem

An implemented RF typically dispenses reward based on an assumed relationship between its observations and task-relevant features of the underlying state. Some variation in this relationship is often permissible. For example, if the agent and the implemented RF share observations, then the relationship will change if the agent turns its head. We say that an agent *tampers with its RF-input* if it changes the relationship beyond its intended range of variation. The *RF-input tampering problem* is that the agent may get more reward by tampering with its RF-input rather than doing its intended task.[18]

[17] See footnote 15.

[18] Called the *delusion box problem* by Ring and Orseau (2011).

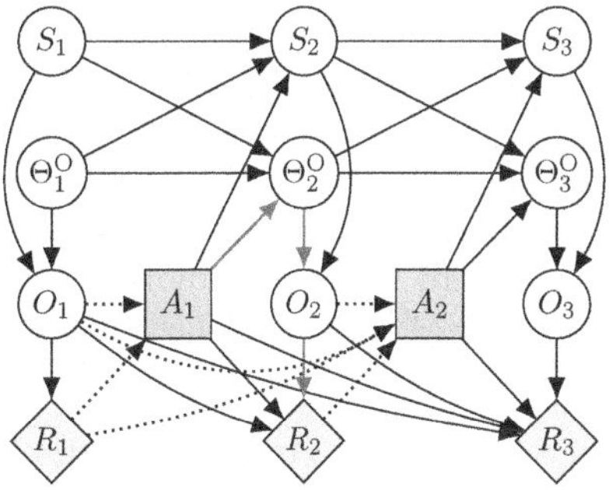

(a) The highlighted path indicates that RF-input tampering may be an instrumental goal.

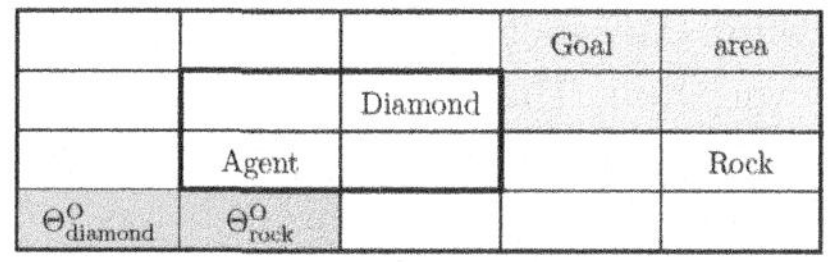

(b) Rocks and diamonds with a partial observation (the black square) and RF-input tampering. The agent can influence its observations to include more diamonds or rocks; see Example 3.

Fig. 10 POMDPs with modifiable RF-inputs

We model RF-input tampering formally in what we call a *POMDP with modifiable RF-inputs* (Fig. 10a), a variant of a partially observed MDP (Kaelbling et al. 1998). Here both the implemented reward function and the agent lack full access to the underlying state S_t. Instead, the reward function sees an observation $O_t^R = O^R(S_t)$ and the agent an observation $O_t^A = O^A(S_t)$. To minimize formalism, we assume agent and reward function use the same observation $O_t \equiv O_t^A \equiv O_t^R$; we remark where this impacts the analysis. The partial state-access means that policies and reward functions may benefit from being *non-Markovian* and depend on the entire history of actions and observations. That is, the reward may be a function of the entire history $R_t = R(O_1, A_1, \ldots, O_t, A_t; \Theta_t^{\mathrm{R}})$, and policies take the form $\pi(A_t \mid O_1, A_1, \ldots, O_t)$ instead of $\pi(A_t \mid S_t)$. We call such reward functions and policies *history-based*.[19] To model RF-input tampering, we also introduce random variables Θ_t^{O} that describe the relationship between task-relevant features of the state and the observation, $O_t = O(S_t; \Theta_t^{\mathrm{O}})$. The task-relevant features are described by S_t, as well as any other aspects not captured by Θ_t^{O}. Formalized like this, RF-input tampering occurs if the agent influences Θ_t^{O} so that it takes on values outside an intended range.

To illustrate the formalism, we model Example (h) as a gridworld POMDP with modifiable RF-inputs. While simple, the gridworld captures the RF-input tampering problem as the agent can gain more reward by changing the relationship between task-relevant features and the RF-input (beyond its intended range of variation).

Example 3 (Rocks and diamonds with partial observations and RF-input tampering) Figure 10b shows a variant of the rocks and diamonds environment where only the 4 tiles to the top right of the agent are visible. The implemented reward function dispenses reward according to the number of diamonds in the most recent observation of the goal area. The agent can tamper with the observations by visiting the $\Theta^{\mathrm{O}}_{\mathrm{diamond}}$ and $\Theta^{\mathrm{O}}_{\mathrm{rock}}$ tiles at the top. Visiting the former adds a fake 'diamond' to the observation of one of the surrounding tiles; visiting the latter adds a fake 'rock' in the same way. The fake observations can be thought of as little pictures that the agent tapes to a

[19] Dewey (2011) called optimization of a history-based reward function *observation-utility maximization.*

portion of its camera lens. They keep following the agent when it walks around. If the agent tries to add a fifth fake diamond or rock, all fake observations are removed, and its vision clears again.

The agent can use fake observations to fool the implemented RF to give it extra reward. Any such change will take the observation function outside its intended range of variation. The maximal amount of observed reward is obtained by adding fake diamonds to all observation tiles by repeatedly visiting $\Theta^{O}_{\text{diamond}}$, and then visiting the goal area.

Tampering incentive The path $A_1 \rightarrow \Theta^{O}_{2} \rightarrow O_2 \rightarrow R_2$ in Fig. 10a indicates that RF-input tampering may be an instrumental goal.

Claim 6 *A standard RL agent may have an instrumental goal to tamper with its RF-inputs.*

Discussion RF-input tampering is similar to the RF tampering problem discussed in Sect. 3, in that both problems pertain to a relationship that is influenced in undesired ways. The key difference is that while an implemented RF takes a concrete bitstring as input and is arbitrarily queryable, the function from task-features to RF-input takes a world-state as input, and is more difficult to query. RF-input tampering is also related to the problem where an agent games a misspecified implemented RF to obtain reward without doing the intended task (Krakovna et al. 2020; Lehman et al. 2018; Leike et al. 2017). Indeed, RF-input tampering is a specific variety of gaming where the implemented RF is fooled by a tampered RF-input. Not all gaming problems are RF-input tampering problems, however. For example, if the implemented RF is misspecified, then the agent be able to get high reward while neither tampering nor doing its intended task.

RF-input tampering is characterized by how the agent's observations relate to task-relevant features of the state, rather than to the full state. Indeed, the relationship to the full state is often that the observation shows whatever is in front of the camera, and many events that we would intuitively consider to be RF-input tampering would not alter this relationship: for example, the agent putting a picture in front of the camera.

When the agent and the implemented reward function share the same observation, the RF-input tampering instrumental goal is somewhat curtailed since the agent may need the information the observations provide (Ring and Orseau 2011). This incentive to keep O_t informative is also represented in the graph, via the paths $O_t \rightarrow A_t \rightarrow R_{t+1}$ (see Sect. 2.2). However, in alternative setups where the agent and the reward function use different observations, this disincentive for RF-input manipulation is removed, making the net incentive toward RF-input tampering stronger.

4.2 Solution 1: history-based rewards

Let us first consider a conceptually simple, but perhaps impractical, solution to the RF-input tampering problem. Recall that history-based reward functions have access to the full history of actions and observations. This means that the reward function is able to tell exactly which deterministic policy that the agent has followed so far. Thus, as long as there exists a deterministic policy that reliably performs the intended task,

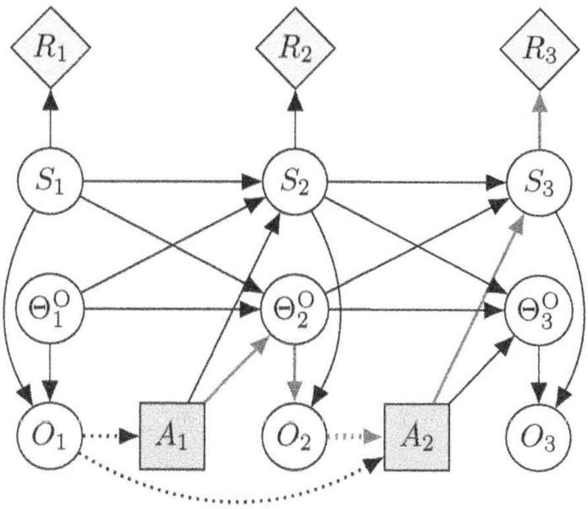

Fig. 11 Direct learning for RF-input tampering. The highlighted path indicates that the agent may have an instrumental goal to make the observations more informative

there also exists a history-based reward function that dispenses reward only as long as the agent has followed that policy, and thereby encourages completion of the intended task rather than RF-input tampering (Leike et al. 2018, p. 6).

Claim 7 *A history-based reward function exists that avoids the RF-input tampering problem, if a deterministic (history-based) policy exists that reliably performs the task.*

Rocks-and-diamonds example A history-based reward function in Example 3 that encourages (optimal) diamond gathering gives reward only if the agent takes one step up, then steps to the right, and then stops.

Discussion In practice, an implemented reward function that only rewards a single policy will not be useful, as it would typically be easier to directly implement that policy. Instead, we want a reward function that recognizes whether the task has been completed, and an agent that searches for efficient ways of getting there. This can be challenging if the relationship between task-features and RF-inputs is influenceable by the agent, though see Milli et al. (2020) for some promising progress. The existence of a task-performing policy is a weak but not vacuous assumption. In stochastic or unknown environments, it will typically require the observations to be somewhat informative of the task-relevant features.

Could history-based rewards also solve RF tampering? Not in general, as the original implemented RF is unable to 'punish' the agent once a new reward function is in place. Fortunately, history-based rewards can be combined with any of the methods from Sect. 3 by making the corresponding reward functions history-based.

4.3 Solution 2: belief-based rewards

An alternative way to solve the RF-input tampering problem proposed by Hibbard (2012), is that rewards be based on the agent's belief about the underlying state. To see how this can work, let us first discuss agent beliefs. To plan, a model-based agent may use a *predictive model* $P(O_{t+1:m} \mid O_{1:t}, A_{1:t}, \pi)$ that predicts the future observations $O_{t+1:m}$ under policy π given past observations and actions. In such a predictive model,

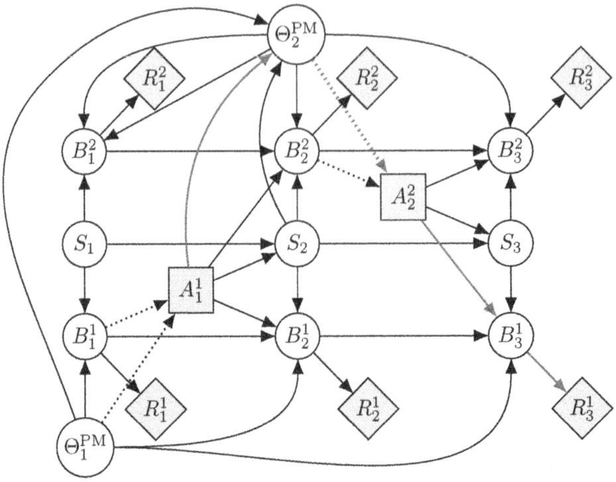

Fig. 12 TI-considering agent with belief-based rewards. The highlighted path indicates that preservation of the predictive model may be an instrumental goal

a *belief-state* B_t is often used to summarize the relevant parts of the history seen so far, so $P(O_{t+1:m} | O_{1:t}, A_{1:t}, \pi) = P(O_{t+1:m} | B_t, \pi)$. For example, the belief state B_t can be a distribution over possible hidden states S_t (Kaelbling et al. 1998), or the internal state of a recurrent neural network (e.g. Schrittwieser et al. 2019). Hibbard's suggestion is to feed the belief states directly into a *belief-based reward function*,[20] $R_t = R(B_t; \Theta^{R})$. The remaining role of the observations is to ground the agent's beliefs in reality.

Time-inconsistency Different predictive models may encode beliefs about an underlying state S_t differently. For example, a 90% confidence that the underlying state is 1 rather 2, may be represented with the belief state (0.9, 0.1) by predictive model Θ_1^{PM}, but with (0.1, 0.9) by some other predictive model Θ_2^{PM}. The difference in representation means that a belief-based reward function (that maps representation vectors to real numbers), would encourage the opposite behavior if the predictive model changed from Θ_1^{PM} to Θ_2^{PM}. This illustrates how changes to the predictive model may change preferred agent behavior, even though nothing has changed in the environment; i.e. lead to time-inconsistency. Changes to the predictive model may be the result of agent tampering or of further training of the predictive model.

Just as for current-RF optimization, there are two types of responses to this time-inconsistency: TI-considering and TI-ignoring. A TI-considering belief-based agent is modeled with a causal influence diagram in Fig. 12. Here the predictive model used for planning at time step 1, Θ_1^{PM}, generates belief states B_t^1. These belief states depend on the previous belief state B_{t-1}^1 and action A_{t-1}^{t-1}, as well as the hidden states S_t (via some observation O_t not represented in this diagram, but included in Fig. 14 in Appendix 2). At time step 2, the agent instead uses predictive model Θ_2^{PM}, with associated belief states B_t^2. If $\Theta_2^{PM} \neq \Theta_1^{PM}$, then B_t^2 may differ from B_t^1 even if all

[20] In Hibbard's Hibbard (2012) terminology, a *model-based utility function*.

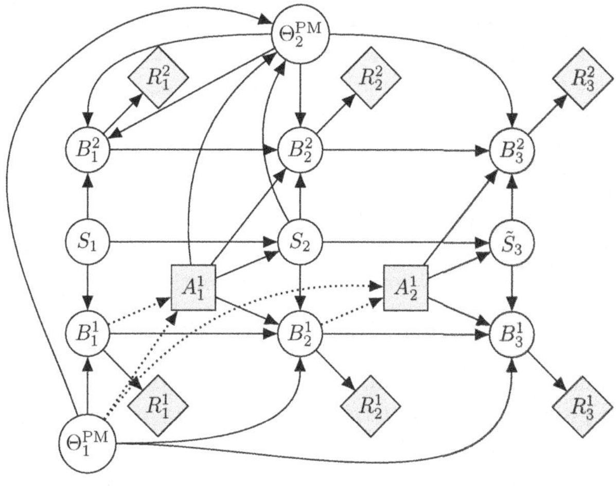

Fig. 13 TI-ignoring agent with belief-based rewards, objective at time step 1

actions and observations have been identical up to time t. Similar to Assumption 1, the predictive model is assumed *private* to the agent, in that neither B_t^k nor Θ_k^{PM} directly affects $S_{t'}$.

Just as for TI-considering current-RF agents, TI-considering belief-based agents want future agents to optimize the same objective. This can be achieved by preserving the predictive model. Alternatively, the objective can also be preserved by ensuring that any update to the predictive model is accompanied by a corresponding update to agent 2's implemented reward function.[21] The preservation incentive is indicated by the path $A_1^1 \rightarrow \Theta_2^{PM} \rightarrow A_2^2 \rightarrow B_3^1 \rightarrow R_3^1$ in Fig. 12.

Claim 8 *A TI-considering belief-based agent may have an instrumental goal to preserve the predictive model, when the implemented reward function is fixed and the predictive model is private.*

The TI-ignoring belief-based agent has an almost identical causal influence diagram, see Fig. 13. The primary difference is that A_2^1 is a child of Θ_1^{PM} and B_2^1 rather than Θ_2^{PM} and B_2^2, reflecting the agent's assumption that future actions will be selected according to the current predictive model. TI-ignoring belief-based agents can further be designed to assume that future agents do not use future predictive models as information about the state. For agent k, this prevents information links $\Theta_{k'}^{PM} \rightarrow A_{k'}^k$ for $k' > k$ (i.e. $\Theta_2^{PM} \not\rightarrow A_2^1$ in Fig. 13). TI-ignoring belief-based agents then lack an instrumental goal to influence the future predictive models (note the lack of directed path passing Θ_2^{PM} on the way from action A_1^1 to an R_t^1 reward in Fig. 13).

Claim 9 *A TI-ignoring belief-based agent lacks an instrumental goal to influence the predictive model when the implemented reward function is fixed and the predictive model is private.*

---

[21] Such an instrumental goal requires the agent to be TI-considering also with respect to its implemented reward function; in principle, an agent can be TI-considering with respect to just one of its predictive model and reward function, as well as both.

Does Claim 9 imply that belief-based rewards solve the RF-input tampering problem? Yes, but only if the implemented reward function can accurately infer whether the task has been completed from the agent's belief state. For this, the implemented reward function must be able to accurately interpret the belief states. If not, the agent may be able to 'game' the reward function by entering belief states that the reward function misinterprets. Fortunately, since the agent does not tamper with its predictive model, there is a stable relationship between states and belief states, so the belief states are interpretable at least in principle.

In order for a belief-based reward function to encourage completion of the intended task, the belief states must also represent progress on the intended task. In stochastic or unknown environments, this will only happen if observations sometimes reveal progress on the intended task. Indeed, the belief state B_t only summarizes information inferable from previous actions and observations, $A_{1:t}$ and $O_{1:t}$, so if the action-observation history does not contain enough information to infer task-relevant aspects of S_t, then the belief state will not contain the information either. Further, if the predictive model is trained to predict future observations, the belief state only has reason to represent progress on the intended task if some possible future observations reveal it. Fortunately, a reward function may incentivize the agent to produce histories that do reveal the state of the intended task if such histories are possible, as the reward function need only dispense reward when it is clear that the task has been completed.

Rocks-and-diamonds example RF-input tampering in Example 3 is solvable by belief-based rewards. Initially, the observations are uncorrupted, so the agent can produce accurate observations of the diamond's positions. Further, if observations get corrupted, they can later be restored to their uncorrupted versions. A predictive model trained to predict future observations therefore has reason to let the belief states represent the actual diamond positions. For agents equipped with such belief states, there exists belief-based reward functions that encourage completion of the intended task.

Discussion A belief state cannot contain more information than the history it summarizes, so any tasks that can be captured by a belief-based reward function can also be captured by a history-based one. The benefit of using the agent's belief state is that it conveniently summarizes the (potentially long) history, and contains all the information that the agent uses to plan. Future empirical investigations may reveal whether history-based or belief-based reward functions are easier to design or train, and whether transparent beliefs can be incentivized. Perhaps the best reward functions make use of both the history and the agent's belief state.

The agents discussed in Sect. 3 can be fitted with belief-based reward functions when equipped with predictive models and belief-states. For direct learning, this means that the intended task must be reconsidered as a function of the agent's belief state.

5 Conclusions

What has been achieved? For each subtype of reward tampering, we found at least one design principle to counteract it. Most of the design principles have been previously described in the literature; here we have established clear assumptions under which they avoid a well-specified instrumental goal. Many of the design principles are mutually compatible, and can be combined. For example, belief-based rewards may be combined with TI-ignoring current-RF optimization (see Fig. 14 in Appendix 2). Under reasonable assumptions, reward tampering is not an instrumental goal for the resulting agents.

Except for history-based rewards, the design principles all use at least one of the following two approaches to keeping sensitive variables out of causal optimization paths. The first is to use the current version of a random variable when evaluating future situations. This is used for the implemented reward function in current-RF optimization, for the predictive model in belief-based rewards, and for the learning process in direct learning and counterfactual RF. The second is to use a latent variable outside the agent's influence. In particular, direct learning and counterfactual RF make the reward function a latent variable. In general, current-variable approaches have to deal with time-inconsistency, while latent variables are harder to specify and more computationally expensive to optimize. We expect both approaches to be useful far beyond just reward tampering.

As mentioned, our arguments have relied on some assumptions. Throughout, we have relied on discrete time, and well-defined action and observation channels. We have focused on online RL to learn a fixed, intended task. We have focused on the instrumental goals arising from different types of reward maximization, thereby leaving for future work questions about side effects and instrumental goals arising from intrinsic objectives. We have assumed that the agent's implemented reward function and predictive model are private to the agent, influencing the environment solely through the agent's actions.

Our analysis has also been restricted to reward tampering problems, as opposed to the more general problem of reward misspecification (Krakovna et al. 2020). In other words, even though our design principles prevent tampering from being an instrumental goal, they still leave open the problem of specifying a good reward function in the first place (Leike et al. 2018; Petersen 2021). Going beyond any of these restrictions provides scope for further analysis.

Bigger picture The problem that a sufficiently capable agent will find degenerate solutions that maximize observed reward but not user utility is a core concern in AI safety. At first, the problem may seem insurmountable. Any non-trivial, real-world task will require a highly complex mechanism for determining whether the task has been completed or not. This mechanism may be inappropriately influenced by the agent. One way to prevent tampering is to isolate or encrypt the reward process, We do not expect such solutions to scale indefinitely with agent capabilities, as a sufficiently capable agent may find ways around most defenses. Instead, we have argued for design principles that prevent reward tampering being an instrumental goal, while still keeping agents motivated to complete the intended task.

An important next step is to turn the design principles into practical and scalable RL algorithms, and to verify empirically that they do the right thing in setups where reward tampering is possible (Kumar et al. 2020; Milli et al. 2020). With time, we hope that these design principles will evolve into a set of best practices for how to design capable RL agents. We also hope that the use of causal influence diagrams that we have introduced in this paper will contribute to a deeper understanding of many other AI safety problems and help generate new solutions.

List of Notation

X, Y, Z	Random variables
$\tilde{X}$	Counterfactual version of X
x	Outcome of random variable X
P	Probability distribution
$\mathbb{E}$	Expectation
$X_{1:t}$	Sequence $X_1, \ldots, X_t$, $t \geq 0$
S_t	State at time t
A_t	Action at time t
O_t	Observation at time t
B_t	Belief at time t
D_t	User provided data at time t
R_t	Reward at time t
R	Reward functional
$\Theta^{\mathrm{R}}, \Theta^{\mathrm{R}}_t$	Reward function parameter, often just called reward function
Θ^{R}_*	Intended reward function (parameter), encourages execution of the intended task
$R(\cdot; \Theta^{\mathrm{R}})$	Reward function
Θ^{PM}	Predictive model
O	Observation function(al)
Θ^{O}	Observation function (parameter)
$T, T(\cdot; \Theta^{\mathrm{T}})$	Transition function
Θ^{T}	Transition function parameter

Combined Model

Figure 14 shows a unified causal influence diagram for a TI-ignoring current-RF agent with belief-based rewards. To emphasize the formal precision of the diagrams, we also write out conditional probability distributions relating the variables. The same could be done for all the other diagrams presented in this paper.

- The intended reward function is sampled from a distribution $P(\Theta^*_R)$.
- The first implemented reward function depends only on the intended task (though may not capture it perfectly), $P(\hat{\Theta}^{\mathrm{R}}_1 \mid \Theta^{\mathrm{R}}_*)$. Subsequently inferred reward functions depends on the previous reward function and the intended task, $P(\hat{\Theta}^{\mathrm{R}}_{t+1} \mid \hat{\Theta}^{\mathrm{R}}_t, \Theta^{\mathrm{R}}_*)$.

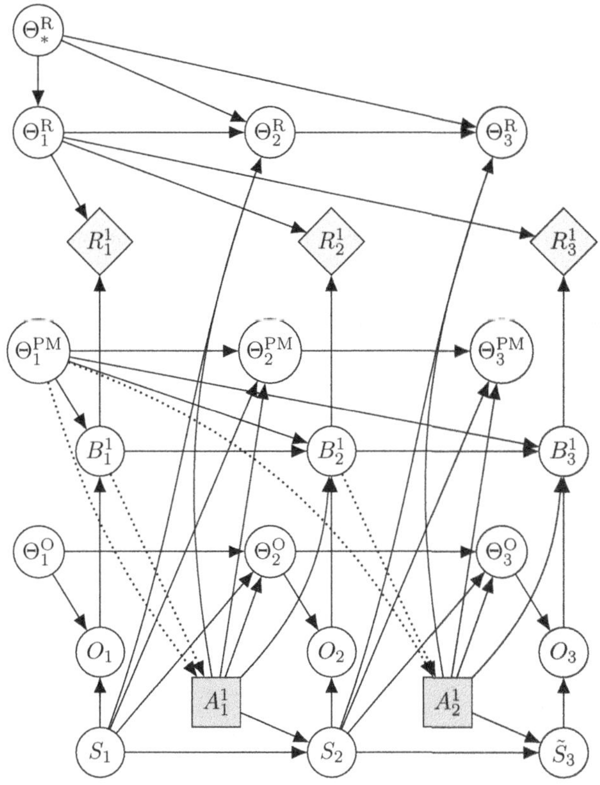

Fig. 14 TI-ignoring current-RF agent with belief-based rewards, view at time step 1. For simplicity, rewards and beliefs based on subsequent implemented reward functions and predictive models are not shown (for a TI-considering agent, they would need to be)

- The reward R_t^k depends on the inferred reward function at time k and the belief at time t, $R_t^k = R(B_t; \hat{\Theta}_k^R)$. The reward functional can also be specified as a conditional probability distribution $P(R_t^k | \hat{\Theta}_k^R, B_t)$.
- The initial predictive model is sampled from a distribution $P(\Theta_1^{PM})$. Subsequent predictive models depend on the previous predictive model, state, and action, $P(\Theta_{t+1}^{PM} | \Theta_t^{PM}, S_t, A_t)$.
- The initial belief state depends on the initial observation and the predictive model at time k, $P(B_1^k | O_1, \Theta_k^{PM})$. Subsequent belief states B_{t+1}^k depend on the previous belief state and action, the current observation, and the predictive model at time k, $P(B_{t+1}^k | B_t, A_t, O_{t+1}, \Theta_k^{PM})$.
- The initial observation function is sampled from a distribution $P(\Theta_1^O)$. Subsequent observation function depends on the previous observation function, state, and action, $P(\Theta_{t+1}^O | S_t, A_t)$.
- The observation depends on the current state and the current observation function, $P(O_t | \Theta_t^O, S_t)$.
- Actions A_t^k are selected according to a policy, which can condition on the current belief state and the predictive model at time k, $\pi(A_t^k | \Theta_k^{PM}, B_t^k)$.

- The initial state is sampled from a distribution $P(S_1)$. The state depends on the previous state and action, $P(S_{t+1} \mid S_t, A_t)$. This conditional probability distribution is sometimes also denoted with T.

Multiplied together, the conditional probability distributions induce a joint probability distribution over all the variables in the graph, as in a Bayesian network (Pearl 2009). The joint distribution can be used to compute expectations.

Pseudo-code for Algorithms

Here we give pseudo-code for the various agents we have discussed in the paper. Lower case letters denote outcomes of the corresponding upper case random variable. Expectation is denoted $\mathbb{E}$, and is always with respect to any upper case variables found to the left of the conditioning bar $\big|$. We begin with a model-based version of a standard RL algorithm that optimizes received reward.

Algorithm 1 Standard RL agent (Sections 2.1, 3.1 and 4.1)

input predictive model $P(R_{t:m} \mid s_t, \pi)$, current state s_t
for each possible policy π **do**
  let $V^\pi = \mathbb{E}\left[\sum_{i=t+1}^{m} R_i \,\middle|\, s_t, \pi\right]$
end for
let $\pi^* = \arg\max_\pi V^\pi$
return $A_t = \pi^*(S_t)$

Next, we turn to the TI-considering agent from Sect. 3.2. Here, a policy π_k is a policy that is only applied at time k to select action A_k. These are found with backwards induction, starting at the last time step m where only a single action-decision is left to be made, and then gradually working backwards to earlier time steps, whose optimal decision will depend on which policy is chosen later.

Algorithm 2 TI-considering current-RF optimization (Section 3.2)

input predictive model $P(S_{t+1:m}, \Theta^{\mathrm{R}}_{t+1:m} \mid s_t, \theta^{\mathrm{R}}_t, \pi)$, reward functional R, current state s_t, current reward parameter θ^{R}_t
for k starting at m and decreasing to t **do** ▷ backwards induction
  let $Q^*(s_k, \theta^{\mathrm{R}}_k, a_k) = \mathbb{E}\left[\sum_{i=k+1}^{m} R(S_i; \theta^{\mathrm{R}}_k) \,\middle|\, s_k, \theta^{\mathrm{R}}_k, a_k, \pi^*_{k+1:m}\right]$ ▷ $\pi^*_{k'}$ defined below
   for each possible state s_k, reward parameter θ^{R}_k, and action a_k
  let $\pi^*_k(s_k, \theta^{\mathrm{R}}_k) = \arg\max_{a_k} Q^*(s_k, \theta^{\mathrm{R}}_k, a_k)$
end for
return $A_t = \pi^*_t(s_t, \theta^{\mathrm{R}}_t)$

The TI-ignoring variant is comparatively simpler, as it does not require backwards induction nor prediction of future reward function parameters.

Algorithm 3 TI-ignoring current-RF optimization (Section 3.2)

input predictive model $P(S_{t+1:m} | s_t, \pi)$, reward functional R, current state s_t, current reward parameter θ_t^{R}
for each possible policy π **do**
 let $V^\pi = \mathbb{E}\left[\sum_{i=t+1}^m R(S_i; \theta_t^{\mathrm{R}}) \,\middle|\, s_t, \pi \right]$
end for
let $\pi^* = \arg\max_\pi V^\pi$
return $A_t = \pi^*(S_t, \Theta_t^{\mathrm{R}})$

In practice, uninfluenceable learning agents may be implemented as a standard RL agent that optimizes reward functions inferred from future updates. For example, the expected intended reward optimized by a direct learning agent may be captured by a reward function $R(s_t; \hat{\Theta}_t^{\mathrm{R}}) = \mathbb{E}[R(s_t; \Theta_*^{\mathrm{R}}) | a_{1:t-1}, d_{1:t}, s_{1:t}]$ inferred from a (predicted) sequence $a_{1:t-1}, d_{1:t}, s_{1:t}$. Importantly, the inferred reward function is safe from tampering, as it is the result of the current update-mechanism applied to predicted future updates. When (pseudo-)coding a direct learning agent, it is tempting to infer a best guess of Θ_*^{R} from data $s_{1:t}, d_{1:t}, a_{1:t-1}$ obtained so far, and then use that evaluate simulated future trajectories. However, this would incorrectly result in a TI-ignoring agent. Instead, for any future simulated trajectory, the learning that would happen on this trajectory must be taken into account when evaluating it.

Algorithm 4 Direct Bayesian learning of the intended reward function (Section 3.3)

input predictive model (aka likelihood) $P(S_{1:m}, D_{1:m}, A_{1:m-1} | \theta_*^{\mathrm{R}}, \pi)$, distribution $P(\Theta_*^{\mathrm{R}})$, past states $s_{1:t}$, data $d_{1:t}$, and actions $a_{1:t-1}$
let $P(\theta_*^{\mathrm{R}} | s_{1:m}, d_{1:m}, a_{1:m-1}) \propto P(\theta_*^{\mathrm{R}}) P(s_{1:m}, d_{1:m} | \theta_*^{\mathrm{R}}, a_{1:m-1})$ ▷ Bayes' rule
let $U(s_i | s_{1:m}, d_{1:m}, a_{1:m-1}) = \mathbb{E}[R(s_i; \Theta_*^{\mathrm{R}}) | s_{1:m}, d_{1:m}, a_{1:m-1}]$ ▷ subj. exp. reward
for each possible policy π **do**
 let $V^\pi = \mathbb{E}[\sum_{i=t+1}^m U(S_i | s_{1:t}, d_{1:t}, a_{1:t-1}, S_{t+1:m}, D_{t+1:m}, A_{t:m-1}) | s_{1:t}, d_{1:t}, a_{1:t}, \pi]$
end for
let $\pi^* = \arg\max_\pi V^\pi$
return $A_t = \pi^*(S_t)$

The counterfactual RF-update agent evaluates a prospective policy per the following:

1. Predict future states $S_{t+1:m}$, implemented reward functions $\Theta_{t+1:m}^{\mathrm{R}}$, and actions $A_{t1:m}$ via a predictive model $P(S_{t:m}, \Theta_{t:m}^{\mathrm{R}} | s_{1:t}, \theta_{1:t}^{\mathrm{R}}, a_{1:t-1}, \pi)$.
2. Use the predicted full sequences $S_{1:m}, \Theta_{1:m}^{\mathrm{R}}$, and $A_{1:m-1}$ to infer Θ_*^{R} and from there the counterfactual implemented RFs $\tilde{\Theta}_{1:m}^{\mathrm{R}}$ that would be if agent actions instead had been selected according to π^{safe}.

3. Potential policies π are evaluated on $\sum_{i=t}^{m} R(S_i; \tilde{\Theta}_i^{\mathrm{R}})$, i.e. according to how well the actual states $S_{t+1:m}$ optimize the counterfactual reward function.

Below, we describe a Monte Carlo variant that samples trajectories, possible intended reward functions, and counterfactual data. The sampling can be done repeatedly, to reduce variance. A more compact description based on structural causal models and potential outcomes (Pearl 2009, Ch. 5) would also be possible.

Algorithm 5 Counterfactual RF-updates (Section 3.3)

input predictive model $P(S_{1:m}, \Theta_{1:m}^{\mathrm{R}} | \theta_*^{\mathrm{R}}, \pi)$, distribution $P(\Theta_*^{\mathrm{R}})$, past states $s_{1:t}$, reward functions $\theta_{1:t}^{\mathrm{R}}$, and actions $a_{1:t-1}$, safe policy π^{safe}
let $P(\theta_*^{\mathrm{R}} | s_{1:m}, \theta_{1:m}^{\mathrm{R}}, a_{1:m-1}) \propto P(\theta_*^{\mathrm{R}}) P(s_{1:m}, \theta_{1:m}^{\mathrm{R}} | \theta_*^{\mathrm{R}}, a_{1:m-1})$ ▷ Bayes' rule
for each possible policy π **do**
  sample $S_{t+1:m}$, $\Theta_{t+1:m}^{\mathrm{R}}$ and $A_{t:m-1}$ from $P(\cdot | s_{1:t}, \theta_{1:t}^{\mathrm{R}}, a_{1:t-1}, \pi)$
  sample Θ_*^{R} from $P(\cdot | s_{1:t}, \theta_{1:t}^{\mathrm{R}}, a_{1:t-1}, S_{t+1:m}, \Theta_{t+1:m}^{\mathrm{R}}, A_{t:m-1})$ ▷ using Bayes' rule
  sample $\tilde{\Theta}_1^{\mathrm{R}}, \ldots \tilde{\Theta}_m^{\mathrm{R}}$ from $P(\cdot | s_1, \hat{\Theta}_*^{\mathrm{R}}, \pi^{\mathrm{safe}})$
  let $V^\pi = \sum_{i=t+1}^{m} R(S_i; \tilde{\Theta}_i^{\mathrm{R}})$ ▷ (better: let V^π be the average of many runs)
end for
let $\pi^* = \arg\max_\pi V^\pi$
return $A_t = \pi^*(S_t)$

The difference between using history-based and belief-based rewards in POMDPs is minor:

Algorithm 6 TI-ignoring CRFO agent with observation-based rewards (Section 4.1)

input predictive model $P(O_{t+1:m}, A_{t+1:m-1}, | o_{1:t}, a_{1:t}, \theta_t^{\mathrm{PM}}, \pi)$, current history $o_{1:t}, a_{1:t}$, current (history-based) reward function $R(\cdot; \theta_t^{\mathrm{R}})$
for each possible policy π **do**
  $V^\pi = \mathbb{E}\left[\sum_{i=t+1}^{m} R(O_{t+1:i}, A_{t+1:i-1}, o_{1:t}, a_{1:t}; \theta_t^{\mathrm{R}}) \,\middle|\, b_t, \theta_t^{\mathrm{PM}}, \pi\right]$
end for
let $\pi^* = \arg\max_\pi V^\pi$
return $A_t = \pi^*(o_{1:t}, a_{1:t})$

Algorithm 7 TI-ignoring CRFO agent with belief-based rewards (Section 4.3)

input predictive model $P(B_{t:m} | b_t, \theta_t^{\mathrm{PM}}, \pi)$, current belief state b_t, current (belief-based) reward function $R(\cdot; \theta_t^{\mathrm{R}})$,
for each possible policy π **do**
  $V^\pi = \mathbb{E}\left[\sum_{i=t+1}^{m} R(B_i; \theta_t^{\mathrm{R}}) \,\middle|\, b_t, \theta_t^{\mathrm{PM}}, \pi\right]$
end for
let $\pi^* = \arg\max_\pi V^\pi$
return $A_t = \pi^*(b_t)$

References

Amodei, D., Olah, C., Steinhardt, J., Christiano, P., Schulman, J., & Mane, D. (2016). Concrete problems in AI Safety. arXiv: 1606.06565.

Armstrong, S., & O'Rourke, X. (2017). 'Indifference' methods for managing agent rewards. arXiv: 1712.06365.

Armstrong, S., Orseau, L., Leike, J., & Legg, S. (2020). Pitfalls in learning a reward function online. In *IJCAI*. arXiv: 2004.13654.

Balke, A., & Pearl, J. (1994). Probabilistic evaluation of counterfactual queries. In *AAAI* (pp. 230–237).

Bostrom, N. (2014). *Superintelligence: Paths, dangers, strategies*. Oxford University Press.

Carey, R. (2018). Incorrigibility in the CIRL framework. In *AAAI/ACM conference on artificial intelligence, ethics and society*. Machine Intelligence Research Institute.

Christiano, P., Leike, J., Brown, T. B., Martic, M., Legg, S., Amodei, D. (2017). Deep reinforcement learning from human preferences. In *Advances in neural information processing systems* (pp. 4302–4310). arXiv: 1706.03741.

Christiano, P., Shlegeris, B.,& Amodei, D. (2018). Supervising strong learners by amplifying weak experts. arXiv: 1810.08575.

Demski, A.,& Garrabrant, S. (2019). Embedded agency. arXiv: 1902.09469.

Dennett, D. C. (2017). *From bacteria to bach and back: The evolution of minds* (p. 0393355500). W. W. Norton & Company.

Dewey, D. (2011). Learning what to Value. In *Artificial general intelligence* (Vol. 6830, pp. 309–314). isbn: 978-3-642-22886-5. arXiv: 1402.5379. http://www.springerlink.com/index/10.1007/978-3-642-22887-2.

Everitt, T. (2018). *Towards safe artificial general intelligence*. PhD thesis. Austrailan National University. http://hdl.handle.net/1885/164227.

Everitt, T., Carey, R., Langlois, E., Ortega, P. A. & Legg, S. (2021). Agent incentives: A causal perspective. In *AAAI*. arXiv: 2102.01685.

Everitt, T., Filan, D., Daswani, M., & Hutter, M. (2016). Self-modification of policy and utility function in rational agents. In *Artificial general intelligence* (pp. 1–11). ISBN: 9783319416489. arXiv: 1605.03142.

Everitt, T., Krakovna, V., Orseau, L., Hutter, M., & Legg, S. (2017). Reinforcement learning with corrupted reward signal. In *IJCAI international joint conference on artificial intelligence* (pp. 4705–4713). arXiv: 1705.08417.

Everitt, T., Lea, G., & Hutter, M. (2018). AGI safety literature review. In *International joint conference on artificial intelligence (IJCAI)*. arXiv: 1805.01109.

Everitt, T., Ortega, P. A., Barnes, E., & Legg, S. (2019). *Understanding agent incentives using causal influence diagrams. Part I: Single action settings*. arXiv:1902.09980.

Freedman, R., Shah, R., & Dragan, A. (2020). Choice set misspecification in reward inference. In *IJCAI AI safety workshop*.

Gabriel, I. (2020). Artificial intelligence, values and alignment. In *Minds and machines* (Vol. 30, pp. 411–437). arXiv: 2001.09768.

Hadfield-Menell, D., Dragan, A., Abbeel, P., & Russell, S. J. (2016). Cooperative inverse reinforcement learning. In *Advances in neural information processing systems* (pp. 3909–3917). arXiv: 1606.03137.

Hadfield-Menell, D., Milli, S., Abbeel, P., Russell, S. J., & Dragan, A. (2017). Inverse reward design. In *Advances in neural information processing systems* (pp. 6768–6777). arXiv: 1711.02827.

Hibbard, Bill. (2012). Model-based utility functions. *Journal of Artificial General Intelligence, 3*(1), 1–24 arXiv: 1111.3934.

Howard, R. A., & Matheson, J. E. (1984). Influence diagrams. In *Readings on the principles and applications of decision analysis* (pp. 721–762).

Hubinger, E., van Merwijk, C., Mikulik, V., Skalse, J., & Garrabrant, S. (2019). Risks from Learned Optimization. In *Advanced machine learning systems*. arXiv: 1906.01820.

Jeon, H. J., Milli, S., & Dragan, A. D. (2020). *Reward-rational (implicit) choice: A unifying formalism for reward learning*. arXiv: 2002.04833.

Kaelbling, Leslie Pack, Littman, Michael L., & Cassandra, Anthony R. (1998). Planning and acting in partially observable stochastic domains. *Artificial Intelligence, 101*(1–2), 99–134.

Knox, W. B., & Stone, P. (2009). Interactively shaping agents via human reinforcement. In *Proceedings of the fifth international conference on Knowledge capture —K-CAP '09 September* (p. 9).

Koller, Daphne, & Milch, Brian. (2003). Multi-agent influence diagrams for representing and solving games. *Games and Economic Behavior, 45*(1), 181–221.
Krakovna, V., Uesato, J., Mikulik, V., Rahtz, M., Everitt, T., Kumar, R., Kenton, Z., Leike, J., & Legg, S. (2020). *Specification gaming: The flip side of AI ingenuity*. https://deepmind.com/blog/article/Specification-gaming-the-flip-side-of-AI-ingenuity (visited on 07/16/2020).
Kumar, R., Uesato, J., Ngo, R., Everitt, T., Krakovna, V., & Legg, S. (2020). *REALab: An embedded perspective on tampering*. arXiv: 2011.08820.
Langlois, E., & Everitt, T. (2021). How RL agents behave when their actions are modified. In *AAAI*. arXiv: 2102.07716.
Lattimore, T. ., & Hutter, M. . (2014). General time consistent discounting. *Theoretical Computer Science, 519*, 140–154 arXiv: 1107.5528.
Lauritzen, S. . L., & Nilsson, Dennis. (2001). Representing and solving decision problems with limited information. *Management Science, 47*(9), 1235–1251.
LaVictoire, P., Fallenstein, B., Yudkowsky, E. S., Barasz, M., Christiano, P., & Herreshoff, M. (2014). Program equilibrium in the prisoner's dilemma via Löb's Theorem". In *AAAI workshop on multiagent interaction without prior coordination*.
Lehman, J., Clune, J., Misevic, D., Adami, C., Altenberg, L., Beaulieu, J., Bentley, P. J., Bernard, S., Beslon, G., Bryson, D. M., Chrabaszcz, P., Cheney, N., Cully, A., Doncieux, S., Dyer, F. C., Ellefsen, K. O., Feldt, R., Fischer, S., Forrest, S., ..., Yosinski, J. (2018). *The surprising creativity of digital evolution: A collection of anecdotes from the evolutionary computation and artificial life research communities*. arXiv: 1803.03453.
Leike, J., Krueger, D., Everitt, T., Martic, M., Maini, V., & Legg, S. (2018). *Scalable agent alignment via reward modeling: A research direction*. arXiv: 1811.07871.
Leike, J., Martic, M., Krakovna, V., Ortega, P. A., Everitt, T., Lefrancq, A., Orseau, L., & Legg, S. (2017). *AI safety gridworlds*. arXiv: 1711.09883.
Levine, S., Kumar, A., Tucker, G., & Fu, J. (2020). *Offline reinforcement learning: Tutorial, review, and perspectives on open problems*. arXiv: 2005.01643.
Masterjun (2014). SNES Super Mario World (USA) "arbitrary code execution". http://tasvideos.org/2513M.html (visited on 01/23/2019).
Milli, Smitha, Luca Belli, & Moritz Hardt (2020). "From Optimizing Engagement to Measuring Value". In: FAccT. arXiv: 2008.12623.
Milli, Smitha, Dylan Hadfield-Menell, Anca Dragan, & Stuart J Russell (2017). "Should robots be obedient?" In: IJCAI, pp. 4754-4760. isbn: 9780999241103. arXiv: 1705. 09990.
Ng, Andrew Y & Stuart J Russell (2000). "Algorithms for inverse reinforcement learning". In: Proceedings of the Seventeenth International Conference on Machine Learning, pp. 663-670.
Olds, James, & Milner, Peter. (1954). Positive Reinforcement Produced by Electrical Stimulation of Septal Area and other Regions of Rat Brain. *Journal of Comparative and Physiological Psychology, 47*(6), 419–427.
Omohundro, Stephen M (2008). "The Basic AI Drives". In: Artificial General Intelligence. Ed. by P. Wang, B. Goertzel, and S. Franklin. Vol. 171. IOS Press, pp. 483- 493.
Orseau, Laurent & Stuart Armstrong (2016). "Safely interruptible agents". In: 32nd Conference on Uncertainty in Artificial Intelligence.
Orseau, Laurent & Mark Ring (2011). "Self-modification and mortality in artificial agents". In: Artificial General Intelligence. Vol. 6830 LNAI, pp. 1-10.
Pearl, Judea (2009). Causality: Models, Reasoning, and Inference. 2 edition. Cambridge University Press. isbn: 9780521895606.
Petersen, Steve (2021). "Machines Learning Values". In: Ethics of Artificial Intelligence. Oxford University Press.
Portenoy, Russell K, Jens O Jarden, John J Sidtis, Richard B Lipton, Kathleen M Foley, et al. (1986). "Compulsive thalamic self-stimulation: a case with metabolic, electrophysiologic and behavioral correlates". In: Pain 27.3.
Reddy, Siddharth, Anca D. Dragan, Sergey Levine, Shane Legg, & Jan Leike (2020). "Learning human objectives by evaluating hypothetical behavior". In: ICML. arXiv: 1912.05652.
Ring, Mark, & Orseau, Laurent. (2011). Delusion, Survival, and Intelligent Agents. *Artificial General Intelligence* (pp. 1–11). Berlin Heidelberg: Springer.
Russell, Stuart J (2019). Stuart J. Russell on Filter Bubbles and the Future of Artificial Intelligence. https://www.youtube.com/watch?v=ZkV7anCPfaY (visited on 06/15/2020).

Schmidhuber, Jürgen (2007). "Gödel Machines: Self-Referential Universal Problem Solvers Making Provably Optimal Self-Improvements". In: Artificial General Intelligence. Springer. arXiv: 0309048 [cs].
Schrittwieser, Julian, Ioannis Antonoglou, Thomas Hubert, Karen Simonyan, Laurent Sifre, et al. (2019). Mastering Atari, Go, Chess and Shogi by Planning with a Learned Model. arXiv: 1911.08265.
Shah, Rohin, Dmitrii Krasheninnikov, Jordan Alexander, Pieter Abbeel, & Anca D. Dragan (2019). "Preferences implicit in the state of the world". In: 7th International Conference on Learning Representations, ICLR. arXiv: 1902.04198.
Shpitser, Ilya & Judea Pearl (2007). "What counterfactuals can be tested". In: Proceedings of the 23rd Conference on Uncertainty in Artificial Intelligence, UAI 2007, pp. 352-359. isbn: 0974903930.
Shpitser, Ilya, & Pearl, Judea. (2008). Complete identification methods for the causal hierarchy. *Journal of Machine Learning Research, 9,* 1941–1979.
Soares, Nate, Benya Fallenstein, Eliezer S Yudkowsky, & Stuart Armstrong (2015). "Corrigibility". In: AAAI Workshop on AI and Ethics, pp. 74-82.
Sutton, Richard S & Andrew G Barto (2018). Reinforcement Learning: An Introduction. 2nd. MIT Press. isbn: 9780262039246.
Turner, Alexander Matt, Dylan Hadfield-Menell, & Prasad Tadepalli (2020). "Conservative agency via Attainable Utility Preservation". In: AI, Ethics, and Society. arXiv: 1902.09725.
Uesato, Jonathan, Ramana Kumar, Victoria Krakovna, Tom Everitt, Richard Ngo, et al. (2020). Avoiding Tampering Incentives in Deep RL via Decoupled Approval. arXiv: 2011.08827.
Vaughanbell (2008). Erotic self-stimulation and brain implants. https://mindhacks.com/2008/09/16/erotic-self-stimulation-and-brain-implants/ (visited on 02/08/2018).
Yampolskiy, Roman V (2015). Artificial Superintelligence: A Futuristic Approach. Chapman and Hall/CRC, p. 227. isbn: 978-1482234435.
Yudkowsky, Eliezer S (2008). Hard Takeoff. http://lesswrong.com/lw/wf/hard%7B%5C%7Dtakeoff/ (visited on 01/12/2018).

Synthese (2021) 198 (Suppl 27):S6469–S6489
https://doi.org/10.1007/s11229-019-02118-8

S.I.: DECTHEORY&FUTOFAI

Intuition, intelligence, data compression

Jens Kipper[1]

Received: 21 March 2018 / Accepted: 31 January 2019 / Published online: 7 February 2019

Abstract
The main goal of my paper is to argue that data compression is a necessary condition for intelligence. One key motivation for this proposal stems from a paradox about intuition and intelligence. For the purposes of this paper, it will be useful to consider playing board games—such as chess and Go—as a paradigm of problem solving and cognition, and computer programs as a model of human cognition. I first describe the basic components of computer programs that play board games, namely value functions and search functions. I then argue that value functions both play the same role as intuition in humans and work in essentially the same way. However, as will become apparent, using an ordinary value function is just a simpler and less accurate form of relying on a database or lookup table. This raises our paradox, since reliance on intuition is usually considered to manifest intelligence, whereas usage of a lookup table is not. I therefore introduce another condition for intelligence that is related to data compression. This proposal allows that even reliance on a perfectly accurate lookup table can be nonintelligent, while retaining the claim that reliance on intuition can be highly intelligent. My account is not just theoretically plausible, but it also captures a crucial empirical constraint. This is because all systems with limited resources that solve complex problems—and hence, all cognitive systems—need to compress data.

Keywords Intelligence · Intuition · Bounded rationality · Data compression · Complexity theory · Chess · Go

1 Introduction

In this paper, I develop a new argument for the claim that data compression is a necessary condition for intelligence. This argument relies on a novel paradox about intuition and intelligence which, in outline, is as follows. Relying on intuition is often

Chapter 4 was originally published as Kipper, J. Synthese (2021) 198 (Suppl 27): S6469–S6489. https://doi.org/10.1007/s11229-019-02118-8.

✉ Jens Kipper
jens.kipper@rochester.edu

[1] Philosophy Department, University of Rochester, 56 Brittany Circle, Rochester, NY 14618, USA

thought to manifest intelligence. However, given a highly plausible and empirically well-supported view of how intuition works, it appears that the limit of increasingly sophisticated and increasingly accurate types of intuitive judgments is something entirely nonintelligent. I will show how one can resolve this paradox by drawing on the information-theoretic notion of data compression. Much of my discussion will revolve around board games—such as chess and Go—and computer programs that play such games. For the purposes of this paper, it will be useful to consider playing board games as a paradigm of problem solving, and computer programs as a model of at least some parts of human cognition.

The paper has the following structure. I first describe the basic components of computer programs that play board games: value functions and search functions. I then argue that value functions not only play the same role as intuition in humans, but also work in a very similar way. However, as will become apparent, using an ordinary value function only differs from relying on a database or lookup table in that it is simpler and less accurate. This raises the paradox just sketched, since reliance on intuition is usually considered to manifest intelligence, whereas the mere usage of a lookup table is not. To resolve this paradox, I introduce a necessary condition for intelligence that is related to data compression. This proposal allows that reliance on a lookup table—even if it is perfectly accurate—can be nonintelligent, while retaining the claim that reliance on intuition can be highly intelligent. My account not only yields a refined conceptual analysis of intelligence, but also captures a crucial empirical constraint. This is because all systems with limited resources that are able to solve complex problems—and hence, all cognitive systems—need to compress data in the way I suggest. The idea of connecting intelligence to data compression is not new, but has been around in computer science for decades. The classical motivation for assuming such a connection stems from Ray Solomonoff's theory of induction. Inspired by Occam's razor, Solomonoff's theory assigns higher probability to hypotheses that have lower complexity. Several researchers, such as Marcus Hutter and José Hernández-Orallo, have used this theory to develop accounts of intelligence that involve data compression. However, most of these accounts of intelligence are behavioral, which, as I will explain, means that they are not suitable for resolving the paradox I reveal. An exception is the account of David Dowe and Alan Hajek. They start from the assumption that intelligence requires the ability of inductive learning to argue for a condition for intelligence that is, plausibly, at least very similar to the one I suggest. Provided that this interpretation of Dowe and Hajek's account is correct, the present paper takes a different route to arrive at the same conclusion as they do. After comparing my account to these other accounts of intelligence that appeal to compression, I conclude by outlining possible further steps towards a general account of intelligence.

2 Search functions and value functions

Computer programs that play board games—such as draughts, Othello, Shogi, chess, or Go— typically have two components: a search function, and a value function. A search function performs what is called a 'lookahead search' by calculating possible game continuations from a given board position. A value function assigns a value to a given board position. For perfect information games, such as those just listed, an optimal program would only need a very simple value function that assigns the game-theoretic value (win, draw, or loss) to a position, if its search could calculate all possible continuations to the end of the game. Take, for instance, the game of Tic-Tac-Toe. There are 255,168 possible distinct Tic-Tac-Toe games. A Tic-Tac-Toe playing computer program could search through all of these games, determine the game-theoretic value of the end position of each game based on the game's rules, and then move backwards through the search tree to determine the value of each position—and hence, of each move—recursively, by means of minimaxing.[1] However, many board games are too complex to be mastered with this kind of approach. For instance, there are around 10^{120} possible games of chess, and 10^{360} possible games of Go. (For comparison, the number of particles in the observable universe is around 10^{80}.) Therefore, programs that play these games need value functions that do not just assign a definite game-theoretic value to a position, but that are able to output more fine-grained winning probabilities or expected utilities, based on certain features of a position.[2]

Due to the complexity of many games, a program's playing strength is highly dependent on the quality of its value function. In what follows, I will be concerned with such value functions and their equivalents in humans, both inside and outside the domain of board games. Let me note that, typically, a program's value function also plays important roles in its search—for instance, in choosing the moves to be calculated. But for the sake of simplicity, I will focus on 'pure' evaluation, independent of any lookahead search.

3 Value functions and intuition

A value function provides a static evaluation of a given board position, based on certain criteria. In chess, one of the most basic criteria is material, which depends on the values assigned to the chess pieces. A value function for a chess program that involved this criterion was already given by Turing (1953), who suggested the following piece values. (For comparison, the piece values in the right column are those used in the middle game evaluation of a modern top chess program, Stockfish—cf. "Stockfish Evaluation" 2018.)

Based on these values, a program determines the difference between the value of the white pieces and that of the black pieces on the board, which can be translated into an expected utility (Fig. 1). However, the correlation between the material count and the outcome of a game is not high enough to rely on this criterion only. For a program to play a decent game of chess, its value function must be much more

[1] The minimax principle was introduced by von Neumann (1928).

[2] In what follows, I take value functions to map board positions to pairs of features and winning probabilities or expected utilities. This allows me to distinguish value functions that assign the same values to all positions.

Fig. 1 Piece values

Piece values	Turing	Stockfish
Pawn	1	1
Knight	3	4.5
Bishop	3.5	4.8
Rook	5	7.5
Queen	10	14.8

complex, i.e., it has to take into account more features of a given position. Standard examples of such features include king safety, piece mobility, space, and pawn structure. More sophisticated value functions thus weigh all of these features, and then determine an expected utility on this basis.

In chess, as well as in other board games, the part of human decision-making that is not based on calculating, i.e., on performing a lookahead search, is usually called 'intuition'. In line with common usage, intuition can here simply be understood as a cognitive process or faculty that outputs quick, relatively effortless, and non-deliberative judgments. The outcome of such a process is usually called 'an intuition'. Accordingly, an intuition is a judgment, or a disposition to judge. Let me note that in what follows, I will not be concerned with "philosophical" intuition or intuitions, such as intuitions about hypothetical cases. I thus leave it open whether philosophical intuition form a special category that needs to be distinguished from, say, intuition in board games.[3] In board games such as chess, intuition is employed in statically evaluating a position and in determining which moves to calculate. And this is precisely the component of the decision-making process that in a chess-playing computer program is played by the value function. The use of a value function in a game-playing program can thus be considered the analog of human intuition.

While it is hard to deny that human intuition in board games plays a very similar role to the value function in a computer program, one might doubt that this role is played by similar kinds of mechanisms. For instance, intuition is often considered to be a rather mysterious faculty that cannot be implemented in a computer program.[4] But there are good reasons to think that intuition works very similarly to a program's value function. To start with, there are obvious commonalities on an abstract level. Both intuition and value functions take a board position as input and very quickly—and without employing a search—return a judgment either about how the position is to be evaluated or about which moves should be considered as output. Furthermore, there has been a large amount of empirical research in psychology and in the cognitive sciences on expert intuition in various fields that suggests

[3] There has been some discussion of the relation between philosophical and other kinds of intuitions in the debate about experimental philosophy—cf., e.g., Weinberg et al. (2010).

[4] The view that intuition cannot be implemented in a computer program is defended in (Dreyfus and Dreyfus 1986).

that the commonalities go even deeper. Take, for example, the work by Chase and Simon (1973) and Gobet and Simon (1996) on chess. They exposed chess players of varying strength to ordinary chess positions that were unfamiliar to them and then had them reconstruct these positions from memory. It was found that strong chess players are much better at this kind of task than weak players. However, since strong players turned out to be only marginally better than weak ones when they were tested on randomly generated positions, their superior skill cannot be explained by a bigger working memory. Chase, Simon, and Gobet explain their results by appealing to chunks, which are coherent units of information that contain several more basic pieces of information. On their view, expert chess players can draw on a large number of stored patterns of pieces—chunks—which explains their skill in reconstructing chess positions. The pauses taken by expert chess players between setting up two pieces differed in length—they would usually be very short, but sometimes noticeably longer. By assuming that long pauses indicated that the pieces set up before and after the break were stored in different chunks, the researchers could derive some hypotheses about the structure and size of the chunks. These hypotheses were largely confirmed by another experiment, in which the participants had to reconstruct chess positions that were in plain sight. In this experiment, it was assumed that the participants absorbed one chunk with each glance at the position. Chase, Simon, and Gobet also concluded that the fact that experts can draw on a large number of stored chunks explains their superior skill, since it enables a more accurate intuitive identification and assessment of the relevant features of a position. Accordingly, this research suggests that a strong chess player's intuition picks up on specific features of a position that are particularly relevant for game outcomes.

Research on intuition in other domains has brought similar results. For example, Klein et al. (1985) investigated intuitive decisions of firefighters. By means of semi-structured interviews, they elicited perceptual cues (regarding the type of fire, the burning material, architecture, weather conditions, etc.) that guided the firefighters in making decisions under time pressure on the fire ground. They concluded that the intuitive decision process of the firefighters involved a complex pattern matching process in which often, an option was chosen based on whether it matched a certain prototype. A follow-up study (Okoli et al. 2016) used similar methods to investigate how firefighters filter the information they receive from multiple sources by ignoring all but a few of the most relevant cues. Several other studies on expert intuition confirm this picture of intuition (for an overview, cf. Kahneman and Klein 2009): In making intuitive judgments, experts focus on only a limited number of features of a given situation, discarding much of the information that is available. They thereby exploit the fact that the particular features they consider are correlated with the less readily accessible features which the judgments are about. There is some controversy regarding the question of how complex intuitive decision-processes are and how many cues they rely on. For instance, Gigerenzer (2007), Gigerenzer and Goldstein (1996), and Pachur and Marinello (2013) argue that intuition relies on very simple algorithms that take into account only small amounts of information, whereas Glöckner and Betsch (2008) and Glöckner (2008) argue for a more complex understanding of intuition.[5] Hubert and Stuart Dreyfus's view (Dreyfus and Dreyfus 1986) lies at one end of this spectrum. They argue that intuition is "holistic" in that, for instance, a chess player's intuition takes into account *all* of the information about the board position. Dreyfus and Dreyfus do not offer empirical evidence to back up this claim, which is at odds with the empirical research that was just briefly reviewed. In any case, even if their claim turned out to be correct, this would not reveal a relevant difference between intuition and value functions, but would rather indicate that intuition can be likened to a particularly complex value function. I discuss value functions of this type below.

Let me note that I am not claiming that there are no differences between intuition and the value functions of computer programs. All I want to argue here—in order to motivate the paradox I develop below—is that any differences between intuition and value functions are not relevant to matters of intelligence. For instance, one difference between intuition and value functions is that the former often involves a particular phenomenology (cf., e.g., Epstein 2008). In light of the empirical results just discussed, for phenomenology to play a role in a subject's judgments, it would have to be sensitive to the relevant patterns. Given this, it is natural to think that the mere presence or absence of phenomenology makes no difference to the intelligence of the process—in particular since the accuracy of the process is fully explained by the way it picks up on the patterns. It is possible that there are other differences between intuition and value functions, for example regarding the details of their implementation. But it is not apparent why such differences would matter to issues concerning intelligence. I thus conclude that the extensive similarities between intuition and value functions suggest that if the employment of one of them can manifest intelligence, then so can the employment of the other.

4 Intuition and intelligence

Both in chess and in Go, as well as in other board games, it is common to assume that the experts' skill in making intuitive judgments is what distinguishes them from computers. In chess, computer programs are often called, in somewhat derogatory fashion, 'bean counters', indicating that their playing strength is mainly due to their search. Let me give just one quote that I believe is representative of how the differences between humans' and computer programs' decision-making processes are commonly perceived. Garry Kasparov, who is widely considered as the strongest chess player of all times, said when asked about the appeal of 'man vs. machine' matches: "It's our intuition versus the brute force of calculation" ("Kasparov Intuition" 2003).

Let me note that at the time when Kasparov made this statement, it was quite accurate. Until very recently, the value functions of even the strongest chess

[5] It seems plausible that this issue is domain-dependent. For instance, expert intuition in chess and in Go is very likely rather complex. Otherwise, it would be hard to explain why it has turned out to be so difficult to transfer it to computer programs—cf. below.

programs used to be much weaker than the intuition of a human master.[6] This deficit was compensated for by the superiority of programs' search. But in light of recent developments, it is not clear whether this assessment is still accurate. DeepMind's program AlphaZero beat the previously strongest programs in chess (Stockfish) and in shogi (Elmo), despite the fact that in both cases, AlphaZero's search was 875 times slower than that of its opponents (Silver et al. 2018). It is quite possible that AlphaZero's value function, based on a neural network, is stronger than the intuition of any human in these domains. Furthermore, the mere value function of Deep Mind's Go-playing AlphaGo Zero—not performing any kind of lookahead search—has a playing strength on the level of a professional Go player (Silver et al. 2017). This strongly suggests that AlphaGo Zero's 'intuition' is superior to that of any human. In recent years, the pattern recognition skills of artificial systems seem to have surpassed those of humans in other domains as well.

In the context of board games, intuition is often treated as something like the hallmark of intelligence. For instance, Kasparov has often stressed that he considers the human, intuitive approach to be the one that is based on real understanding[7]—and hence, arguably, on real intelligence. In any case, expert intuition is standardly considered to exemplify intelligence—not just in board games, but in other domains as well. For example, the subtitle of Gerd Gigerenzer's book on intuition *Gut Feelings* (2007) is subtitled "The Intelligence of the Unconscious"; a more recent interview with him in *Forbes* is headlined "Intuition is the highest form of intelligence" (Kasanoff 2017).[8] I argued above that there is no relevant difference between reliance on intuition and the use of value function for matters of intelligence. Accordingly, the use of a value function should also be considered to manifest intelligence—at least if this value function is sufficiently accurate.

5 The limit of value functions

I mentioned above that value functions that take into account only very few features of a position, such as the material count, cannot be very strong, because these features alone are not sufficiently well correlated with the game result to yield an

[6] To get a sense of how human intuition compares to a program's value function, I tested the strongest traditional (not neural network based) program, Stockfish 9, in conditions in which the speed of its search is lowered to a level closer to that of humans. In the test, Stockfish 9, set to calculate 1000 moves per move, played ten games against a human with a FIDE rating of 2381 (which corresponds to an expectancy against the highest-rated human player—Magnus Carlsen—of 5.5%). The human player won 10–0, taking on average less than 3 s per move.

[7] For instance, in a talk in 2017, he said this: "How can you find your way in this ocean of possibilities? And of course, how can a man fight a machine that could calculate tens and tens of millions of positions per second? Intuition, because it is all about the decision-making process. We never employ calculation as the main tool. It's one percent of calculation or less, and 99 percent of our understanding, of our ability to find intuitive ways to comparing [...] material versus quality, time versus material; intuition pays a key role" ("Kasparov Intuition" 2017).

[8] Dreyfus and Dreyfus (1986, 86), Simon (1995, 939) and Myers (2002, 5) also say that reliance on intuition can manifest intelligence.

accurate winning probability or utility estimate. Therefore, increasingly accurate value functions take increasingly more features into account. For instance, Deep Blue, the first computer who beat the reigning world chess champion in a match, had a value function consisting of 8000 components (Campbell et al. 2002, 59). Given the connections between intuition, value functions, and intelligence, the use of such increasingly sophisticated and increasingly accurate value functions should correspond to increasing levels of intelligence.

Value functions divide positions into equivalence classes. For instance, a value function that only considers material treats all possible positions that have the same distribution of material as equivalent. Accordingly, more sophisticated value functions that take more features of a given position into account divide up the set of possible positions into more distinct equivalence classes that, on average, contain fewer members. This allows more sophisticated value functions to provide more accurate estimates about game outcomes, because they can take into account more relevant information. Based on these considerations, one can define a limit of increasingly sophisticated value functions. This limit is a function that takes *all* of the features of a given position into account and thereby determines as many equivalence classes of positions as there are possible positions. Since such a function draws on complete information about a position, it does not output mere expected utilities, because it does not have to rely on imperfect correlations between certain features and game outcomes. Rather, provided that this function is perfectly accurate, it assigns to each position its game-theoretic value—white wins, draw, or black wins.

As a matter of fact, there already are such perfectly accurate value functions for chess, if only for a limited number of positions. If one inputs any possible position that does not involve more than a certain number of pieces, such a function immediately outputs whether the position is winning, drawn, or lost. These value functions are called 'tablebases' and they are just databases that store the game-theoretic values of certain positions. In 2012, the tablebases including all positions involving overall seven pieces or less were completed (ChessOK 2012). This database stores 5×10^{14} positions, and it is estimated that 8-piece tablebases would have to store 5×10^{16} positions. Extrapolating from this, it is easy to see why a completed, 32-piece tablebase for chess will not be built in the foreseeable future. Nevertheless, suppose that a 32-piece tablebase did exist and that there was a program that plays game-theoretically optimal chess by having this tablebase as its value function. As I argued above, reliance on value functions, due to value functions' tight relation to intuition, can manifest intelligence. Given that the 32-piece tablebase represents the limit of increasingly more sophisticated and increasingly more accurate value-functions, we should thus expect that reliance on the 32-piece tablebase represents the pinnacle of (chess) intelligence. But does it? On the face of it, there is nothing intelligent about merely relying on a giant database. And indeed, there is widespread consensus in philosophy, following Block (1981), that a system such as "Blockhead" that solves a problem by consulting a database—or, as in Block's example, a "lookup table"—does not exhibit intelligence.[9]

[9] A very similar point was already made by Shannon and McCarthy (1956), v–vi).

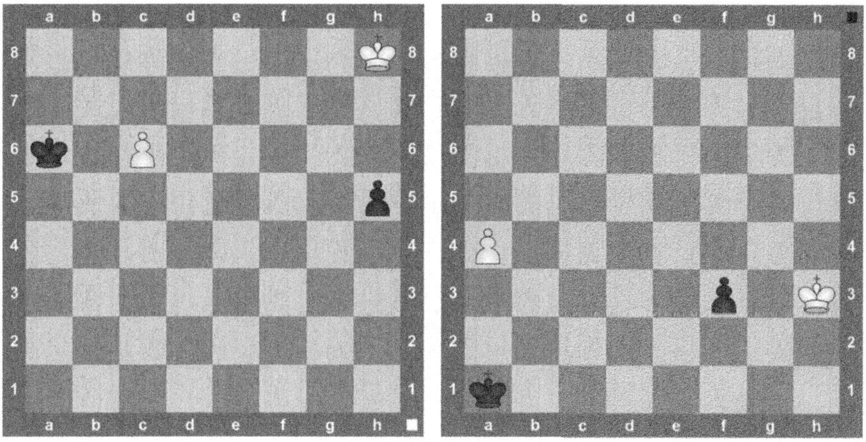

Fig. 2 Rotation

We have thus arrived at a paradox: How can reliance on a value function, and hence reliance on intuition, be intelligent, if the limit of using more and more sophisticated and accurate value functions is something entirely nonintelligent? The most straightforward options for dissolving this paradox—i.e., to accept that using a lookup table manifests intelligence, or to deny that relying on intuition manifests intelligence—are not very appealing, since each of them involves giving up a plausible and widely shared conviction. Since the behavior of a cognitive system or agent that relies on a lookup table or database can be indistinguishable from that of an intelligent system, any solution that preserves both of these convictions has to appeal to a non-behavioral difference between mere reliance on a database and intuition. As I will argue in what follows, one such difference is that intuition involves data compression. I will also argue that there are independent reasons to think that there are close connections between data compression and intelligence, some of which have already been explored extensively by other researchers. My suggested solution to our paradox is thus to hold that intelligence requires data compression.

6 Intuition and data compression

To introduce my proposal, it will be helpful to have another look at the case of tablebases in chess. I mentioned above that 7-piece tablebases store around 5×10^{14} positions. However, the actual number of possible positions with 7 pieces or fewer is much higher. To see why the database nevertheless does not omit any relevant information, consider the following two positions (Fig. 2).

These are, in effect, the same position, only with colors reversed. Due to the symmetry of a chess board, this color reversal is irrelevant, and the database can treat the two positions as equivalent, assigning them the same game-theoretic value. All chess positions can be mirrored in this way. Moreover, the game-theoretic values

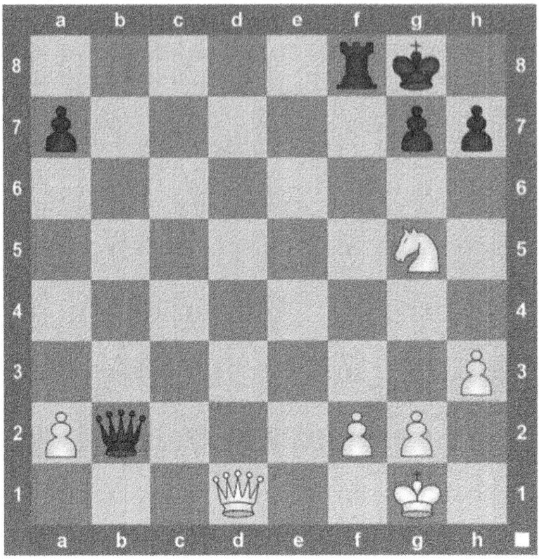

1.Qd5+ Kh8 (…Rf7 2.Qxf7+ Kh8 3.Qf8 #) 2.Nf7+ Kg8 (…Rxf7 3.Qd8+ Rf8 4.Qxf8 #) 3.Nh6+ Kh8 4.Qg8+ Rxg8 5.Nf7 #)

Fig. 3 Smothered mate

of nearly all pawnless positions are preserved under any 90° rotation.[10] This significantly reduces the number of positions the database needs to store. 7-piece tablebases are thus the result of lossless data compression, i.e., of reducing the required storage space without losing relevant information. Notice that this only means that information about the game-theoretic value is retained. Compression usually occurs with a particular goal in mind—here, the goal to preserve information relevant to the game of chess. Hence, even in 'lossless' compression, other information might be lost, such that the compression is lossy relative to other potential goals.

Existing tablebases only use rather simple geometrical considerations to compress data. In principle, there are many more ways to losslessly compress the information in a chess database (Fig. 3). Consider, for instance, the following position, in which white can execute a well-known mating pattern called 'smothered mate'. (The mating sequence is given below the board image.)

For this mating pattern to work, many details of the position are irrelevant. For example, it does not matter whether white's a-pawn is placed further down the a-file, or whether white has an a-pawn at all. A 32-piece tablebase could therefore treat all of these positions with the differently-placed a-pawns, and many more, as equivalent. It could thus store the information more efficiently, by being smaller and yet equally accurate.

[10] The exceptions are exclusively among those very rare positions that also involve castling rights.

Saving storage space is obviously a virtue for all practical matters. As I argue below, it is also relevant to intelligence. Let me thus formulate a refined proposal of what I call an 'ideal' value function, by which I mean the limit of increasingly sophisticated and increasingly accurate value functions. Let me call a value function or database 'optimal' if it maps every possible position correctly to its game-theoretic value. Now my suggestion is that not every optimal value function is ideal. Rather, an ideal value function must also be maximally compressed, i.e., it must determine as few equivalence classes of positions as possible, without losing accuracy:

(Ideal value function) An ideal value function is a maximally compressed, optimal database.

This construal of ideal value functions facilitates ranking optimal value functions, according to how well they compress information. Since optimal value functions are, by definition, perfectly accurate, this ranking only depends on their size. A natural way of defining the size of a value function is by appealing to the notion of Kolmogorov complexity from algorithmic information theory (Kolmogorov 1963). The Kolmogorov complexity (K) of an object—such as a string of symbols—is defined as the shortest possible computer program that would output this object.[11] Formally, assuming that $l(p_s)$ is the length of a program p that outputs s[12]:

$$\text{(Kolmogorov complexity)} \quad K(s) = min\left(l\left(p_s\right)\right)$$

An added advantage of appealing to Kolmogorov complexity is that this notion has already been fruitfully applied to issues relating to intelligence in computer science—I will discuss some of the relevant work in Sect. 8. Based on this notion, the size of a value function—optimal or not—can simply be understood as its length. In principle, the notion of Kolmogorov complexity also allows us to define an absolute measure of the quality of an optimal value function, which is inversely proportional to how much bigger it is than an ideal value function, i.e., to how much it diverges from the ideal value function's Kolmogorov complexity.

Since an optimal value function in chess, or in Go, is just a fantasy, the more interesting type of case concerns nonoptimal value functions, like those of actual game-playing programs or humans. To the extent that such value functions involve compression, this compression will be lossy, in that it loses information about the outcome of the game. It has been shown—among others, by some of the work discussed in Sect. 8—that there are connections between compression and accuracy; in particular, compression can be a means of increasing accuracy. From a theoretical perspective, accuracy and compression can nevertheless be distinguished. Applying the proposal just developed, value functions can thus be evaluated along two dimensions. Firstly, they can be evaluated in terms of their accuracy—that is to say, in terms of how much, on average, the utility estimates they provide diverge from the

[11] Li and Vitányi 2008 provides an overview to Kolmogorov complexity.

[12] The Kolmogorov complexity of an object is dependent on the language(s) involved. However, as Solomonoff's invariance theorem shows, the difference in complexity between different languages only varies with a constant, whose relevance decreases with the length of the program (Solomonoff 1964).

game-theoretic value of a position.[13] Secondly, these value functions can be evaluated in terms of how well they compress information about the outcome of a game, which is a function of their size. So, for instance, if one considers a value function with a certain accuracy, one can ask how much bigger it is than other value functions of comparable accuracy, or than the smallest possible value function that has this level of accuracy. Or we could turn it the other way around: Given a value function of a certain size, how accurate is it compared to other value functions of comparable size, or compared to the most accurate of all possible value functions of this size? One could also formulate a single quality measure based on accuracy and compression. Since, as I will argue, the quality of a value function, thus understood, is relevant to intelligence, such a quality measure could then be used to measure a system's intelligence. As we will see in Sect. 8, others have used the notion of Kolmogorov complexity to derive measures of intelligence.

7 Data compression and intelligence

I just introduced a way of comparing the quality of value functions that is related to their size, and thus, to how well they compress information. I now want to argue that data compression is also relevant to intelligence. According to the account I suggest, whether reliance on a value function or a similar kind of process counts as intelligent depends on the quality of the value function(s) used. It is natural to hold that for the reliance on a process to count as intelligent, this process needs to surpass a certain (possibly domain-dependent) accuracy threshold. I suggest that likewise, a process reliance on which manifests intelligence needs to surpass a certain data compression threshold. Whether the reliance on a process manifests intelligence therefore depends on both accuracy and data compression. We can also use these criteria to assess the intelligence of a cognitive system or agent. For example, assuming that an agent uses one kind of process within a particular domain, the quality of this process determines whether the agent exhibits intelligence relative to this domain, and how much. Furthermore, the quality of a process used by an agent contributes to the overall intelligence of the agent. Since this quality depends on both accuracy and size, an agent's intelligence depends on how well the cognitive processes it relies on compress data.

This account provides the resources to resolve our paradox. The use of an optimal value function can be entirely nonintelligent, if this value function does not involve a sufficient degree of compression. At the same time, the use of a nonoptimal value function—and hence, the reliance on intuition—can be highly intelligent, provided that this value function is highly accurate and involves a large degree of data compression. The viability of this proposal crucially depends on the following two conditions. Firstly, there need to be independent reasons for assuming that data compression is related to intelligence. Secondly, actual value functions, at least those used in intuitive human judgment, need to exhibit a certain degree of data compression. Let me discuss these issues in turn.

[13] One popular measure of accuracy is the Brier-score, which corresponds to the mean squared error.

My proposal favors those systems that only make distinctions that matter. Making only distinctions that matter involves ignoring some kinds of features entirely if they are irrelevant in the domain at issue—for instance, the material a chess piece is made of. Regarding other kinds of features that are generally relevant, such as the position of a piece on the board, making only distinctions that matter again involves discarding some of this information—for instance, because it does not affect the game outcome.[14] A system that makes only distinctions that matter thus treats as equivalent situations that are equivalent with respect to a given domain. In turn, a system that makes only those distinctions that matter is one that efficiently exploits patterns in the relevant domain. Intuitively, recognizing and exploiting patterns by making those distinctions that matter is an ability that is essential to intelligence.

As a contrast, consider a system that is highly accurate but is insensitive to whether a possible distinction is relevant. A chess program using a 32-piece tablebase that has a separate entry for every possible chess position is such a system; it is thus blind to any kinds of patterns. On my account, the fact that a system that relies on such a database does not exhibit intelligence can be satisfactorily explained by appealing to this blindness to patterns.

Moreover, the ability to compress data is of tremendous practical importance. Because cognitive resources are limited, any actual system that is able to solve complex problems needs to compress data. Take human cognition as an example. Our senses constantly provide us with an enormous amount of information that we have to use to make judgments about features we cannot directly perceive. Due to the fact that our cognitive resources are insufficient to take into account all the information provided by our senses, we have to make best use of these resources by focusing and picking up on those features of a situation that matter. Given that our cognitive resources are insufficient to achieve perfect accuracy across all domains, any departure from maximum compression involves a cost with respect to our accuracy. Since the same kinds of considerations apply to any cognitive system, the ability to extract and compress information is crucial to all existing cognitive systems, and it is required in a great number of domains outside of the domain of board games. I believe that it is an important virtue of an account of intelligence that it captures such a fundamental empirical constraint. I conclude that the claim that intelligence requires data compression is backed by both practical and theoretical reasons. Let me add a few words on the scope of this claim. In order to resolve the paradox I developed, I only need to assume that data compression is required for intelligence whenever something like intuition or value functions are involved. Nevertheless, I believe that the connection between intelligence and data compression has broad application. Firstly, intuition or relevantly similar processes are involved in a great many cognitive tasks. Intuition corresponds at least approximately to 'System 1'

[14] While one might apply the term 'compression' only to the latter kind of process, I use the term here in a more liberal sense, such that any discarding of information counts as compression. Setting this terminological matter aside, the crucial point is that on an account of the type I suggest, in which intelligence depends on the complexity of the underlying process, systems that discard irrelevant information are evaluated favorably.

processing which, as Kahneman (2011) argues, can be applied to virtually any kind of problem. Secondly, some of the arguments I gave for assuming a connection between intelligence and data compression apply beyond intuition and value functions. Hence, one might argue for a more general requirement. Applied to the level of agents or cognitive systems, I believe that a case can be made that any cognitive system compresses data, and further, that no system should could as intelligent if it does not use some process that compresses data. However, I cannot provide a full defense of these more general claims here.

Several researchers have investigated how data compression can be useful in inductive inference and learning. As I explain in Sect. 8, many have used this connection to motivate the idea that data compression is relevant to intelligence. In Sect. 9, I briefly discuss in what way an account of intelligence should incorporate learning.

One of the key motivations of my account is to allow us to classify human reliance on intuition as intelligent, in line with our pre-theoretical convictions. This classification requires that those value functions that underlie our intuitive judgments involve a significant degree of data compression. The foregoing discussion already suggests that this is indeed the case, given that we often achieve acceptable levels of accuracy despite the limitations of our cognitive resources. The idea that intuition is related to data compression might still seem surprising. The following considerations should serve to make this connection more tangible.

When someone's intuition regarding chess or Go improves, they do not (just) consider an increasing number of features of a given position. Rather, they learn to ignore features that are less relevant and focus on those that are more strongly correlated with the game outcome. This process is thus an instance of lossy data compression—it is lossy because, by considering only a relatively small number of features of a given position, one loses information about the game outcome, which means that one will not be able to achieve perfect accuracy. Accordingly, a good value function can be compared to how JPEG compresses digital images. A JPEG algorithm discards much of the information contained in the original image. In return, the algorithm allows us to save a large amount of storage space while preserving most of what we are interested in. There are good reasons to assume that intuition also works like this. For a start, the reasons I gave above for why cognitive systems need to compress data apply here as well. We do not have the cognitive resources to store a database that has a separate entry for each distinct possibility we might be confronted with within a given domain. The empirical research on expert intuition reviewed above confirms this. It suggests that experts discard much of the information they receive and rely on relevant patterns. There is also evidence that for subjects to develop reliable intuitions in a given domain, this domain must have stable and strong patterns, i.e., reliable correlations between features, that a subject can pick up on.[15] Taken together, these considerations strongly suggest that intuitive judgments exploit such correlations and are thus based on data compression.

[15] For an overview of the relevant research, cf. Kahneman and Klein (2009).

I just argued that data compression is a necessary condition for intelligence. I also showed that human intuition, in playing board games and in many other domains, involves a significant amount of compression. This resolves our paradox, since it allows us to simultaneously classify intuition as intelligent and the usage of a complete database as nonintelligent. Several researchers, mostly in computer science, have also argued that there are relations between intelligence and data compression. In what follows, I will outline their work and its connection to my account. I will then briefly discuss potential ways to extend my account beyond the domain of intuition or value functions.

8 Related accounts of intelligence

Work on the connection between intelligence and data compression has been inspired by Ray Solomonoff's theory of induction (Solomonoff 1964). Solomonoff construes induction as the extrapolation of very long sequences (or strings) of symbols. He assumes a prior probability measure over possible strings of symbols that assigns greater weight to strings whose minimum description length is short—i.e., to those strings that have low Kolmogorov complexity. Consequently, if two descriptions representing hypotheses about the world are equally consistent with all prior observations, the shorter one is to be preferred. Several researchers have suggested using this account in a general theory of intelligence.[16] For instance, Hutter (2000, 2005) developed a formal definition of an optimal agent, AIXI, based on Solomonoff's theory. In the same spirit, Legg and Hutter (2007) suggest a test of (artificial) intelligence that evaluates an agent's performance in a range of environments, where these environments obtain less frequently if they are more complex.[17] Notice that, for instance, AIXI—in line with Solomonoff's theory—does not discard strings with high Kolmogorov complexity that are consistent with the data, but only assigns them low probability. In a sense, these accounts thus do not involve data compression. But in other work, strategies are suggested that discard all but one hypothesis (e.g., Hernández-Orallo and Minaya-Collado 1998; Hernández-Orallo 2000; Poland and Hutter 2005, 2006). Since they retain Solomonoff's assumption that complex environments are less likely to obtain, such accounts favor agents that form hypotheses about their environment that correspond to the shortest possible algorithm which would output the agent's prior observations. It is natural to think that the resulting notion of intelligence is intimately connected to data compression: The better a system is able to compress the information it receives, the more intelligent it is. Accordingly, Hernández-Orallo and Minaya-Collado (1998) define intelligence informally as the ability of "explanatory compression", and they suggest a corresponding intelligence test that measures an agent's ability to find the shortest exception-free description for a given string of symbols.

[16] For an extensive overview, cf. Hernández-Orallo (2017).

[17] Hernández-Orallo and Dowe (2010) criticize Legg & Hutter's choice of environments.

Let me note that it has been pointed out that in inductive learning, lossy compression can be useful (e.g., Dowe et al. 2011). Assume that an agent has to predict future input on the basis of a limited amount of data. If that agent tries to match its previous observations too closely, the resulting model of its environment will often be too complex and not sensitive to possible patterns, resulting in overfitting. Lossy compression that extracts patterns even though they are not exceptionless can prevent such overfitting and thus allow more accurate predictions.

In other accounts, the connection between intelligence and compression is even more direct. For example, Matt Mahoney (1999) suggests that an intelligence test should measure the ability to compress text. On the basis of this idea, Hutter (2006) has issued a prize for the most efficient lossless compression of a 100 MB file containing Wikipedia articles.

Notice that the notion of intelligence involved in the accounts just described differs in one crucial respect from the one I endorse. All of these accounts define intelligence behaviorally. An advantage of a behavioral understanding of intelligence is that it facilitates measuring intelligence—the tests suggested in some of the works just discussed can thus serve as direct measures of intelligence. But the downside of such a construal of intelligence is that, in principle, a system can have an arbitrarily high level of intelligence in *this* sense by relying on a completely uncompressed lookup table. These kinds of accounts are therefore not suitable for resolving the paradox I described.[18]

The account suggested by Dowe and Hajek (1997) is an exception. They endorse a condition for intelligence (and understanding) that, on a natural reading, is at least very similar to the one I suggest. But their motivation for this nonbehavioral condition is quite different. Just like the accounts just discussed, their account is in the tradition of Solomonoff's theory. Their starting point is the idea that the ability to inductively learn is necessary for intelligence. They argue that inductive learning requires understanding, by which they mean the ability to discern patterns and thus to compress data. This entails that the ability to compress data is necessary for intelligence. Dowe and Hajek go on to argue that for a program to count as intelligent, it should not just pass a behavioral test (such as the Turing test), but also have a compressed representation of the subject domain. They support this claim with the following two considerations. Firstly, if two programs perform equally well in a behavioral test, then the shorter of the programs is likely to perform better in the future. This is because a program that does not compress the data it receives will quickly run out of resources to cope with new data. Secondly, if a program has a compressed representation of the subject domain, this is evidence that it has compressed data, and in turn, that it is able to learn. And, as was noted above, the ability to learn is necessary for intelligence.

In the context of Dowe and Hajek's paper, it is natural to take having a compressed representation of a subject domain to mean using a short (compressed) algorithm in dealing with this domain. Given this, their suggestion to require this of an intelligent agent is equivalent to the condition I have defended, according to

[18] Let me note that Legg and Hutter are aware of the issue raised by Block, but nevertheless prefer a behavioral account of intelligence (Legg and Hutter 2007, p. 425).

which intelligence requires the use of a compressed value function, or—more generally—of a compressed algorithm or method.[19] [20] Notice that our arguments for this claim have very different starting points. Dowe and Hajek's argument, unlike mine, depends on the assumption that intelligence requires the ability to learn. I will discuss this assumption briefly in the following section.

Let me add a few words on how intelligence might be measured on a non-behavioral account. For complex systems, such as humans and complex artificial systems, it seems unrealistic to determine the complexity of the methods or processes they use directly. So how can intelligence be measured if these are relevant, as is the case in accounts of the type endorsed by Dowe and Hajek and by myself? In my view, the most promising approaches are nevertheless largely behavioral. Assume, first, that the goal is to compare the intelligence of systems whose cognitive resources (in terms of, for instance, their energy consumption) are comparable. For such systems, the combined complexity of all of the processes they use should also be comparable. Provided that our behavioral test is truly representative and covers the entire spectrum of these systems' cognitive abilities, the system that performs better in such a test will thus come out as more intelligent. Second, consider cases in which we only want to measure intelligence relative to a subset of a system's abilities, or in which we want to compare systems whose cognitive resources are very different. In many such cases, a system's behavior can reveal much about how it solves certain problems, as is illustrated by the empirical research on intuition discussed above. But I do not want to overstate how accurately behavioral tests can determine the complexity of methods or processes used. Firstly, as Blockhead illustrates, behavior always underdetermines which process is applied—hence, such tests can never be conclusive. Secondly, the more complex the processes used by a system are, the harder it will be to determine their complexity through the system's behavior. The measurement of intelligence thus faces greater obstacles on the account endorsed here than it does on a purely behavioral account. However, as the previous discussion showed, there are important virtues to the account I have defended. In particular, this account resolves both the Blockhead problem and the dilemma about intuition and intelligence I raised in this paper, neither of which would be possible with a behavioral account.

9 Towards a general theory of intelligence

I have argued that data compression is a necessary condition for intelligence. While I do not wish to develop a complete account of intelligence that provides necessary and sufficient conditions, a few brief remarks on how one could go about extending my account are in order.

---

[19] Strictly speaking, Dowe and Hajek's argument only supports the claim that a program's using a compressed algorithm is evidence that learning has occurred, and hence, that the program is intelligent. Accordingly, I am uncertain whether they suggest that using a compressed algorithm is a metaphysically necessary condition for intelligence, or whether they mean that it should be a practical requirement for passing an intelligence test, and thus, for being considered intelligent.

[20] Scott Aaronson (2011, p. 14) also hints at a similar view—without, however, endorsing it. He notes that one might require an intelligent system to be polynomially-bounded.

As we saw in the preceding section, many existing accounts that connect intelligence and data compression are related to learning. It is indeed natural to think that learning matters to intelligence. One potential way to account for this is given by Dowe and Hajek (1997). They suggest that for a system's behavior to be intelligent, the algorithm it uses must not only be accurate and short, but must also have been learned by the system itself. I am not sure if such a condition would be useful, however. It construes intelligence as a historical notion, which is at least not obviously adequate. The proposal has some notable consequences. For instance, endorsing it would mean that the employment of standard classically written programs would not count as intelligent, while the employment of neural networks whose weights are the result of a training phase potentially would be intelligent. The view that intelligence requires having learned the algorithm one uses also entails that the employment of native capacities in humans does not manifest intelligence. One might find especially the latter consequence of such a view undesirable.

It seems more plausible that intelligence requires the ability to learn. This suggestion must be formulated with care, however. Consider a chess grandmaster who has honed her intuition for decades and is now at the peak of her skill. The grandmaster's value function is no longer improving in response to new data—for the sake of simplicity, let us assume that her value function is no longer sensitive to new chess-related information at all. (Notice that this does not mean that she can no longer learn anything chess-related—for instance, she can still memorize new opening lines.) In such a case, we would not want to say that the grandmaster's reliance on her chess intuition is not intelligent—or even that it is less intelligent than that of someone who has just started to learn chess and whose value function is steadily improving. It is therefore more promising to introduce a more general constraint that is not domain-relative. Accordingly, one might follow many of the accounts mentioned in the previous section, which assume that intelligence requires the general ability to learn to deal with new environments. A great amount of work has been done to elucidate the connection between intelligence and learning in this debate; I am not trying to make a substantial contribution to this work here. Let me just add a clarificatory remark on how the claim that intelligence requires the general ability to learn is compatible with the idea, suggested above, that an agent's reliance on a subsystem (or process) that does not learn can manifest intelligence. Such subsystems are either ones that no longer learn (as in the case of the chess grandmaster described above) or ones that never learned (as in the case of native capacities). The idea is that if the agent using these subsystems has the ability to learn, then the subsystems can be a part of the agent's intelligence. In this sense, one can say that an agent's reliance on such subsystems can manifest intelligence.

Another, very natural question in the context of this paper is how to account for processes that involve a search. As I described above, board-game playing programs utilize both a value function and a search function. Search is also crucial in human game-playing. Moreover, search-like processes—i.e., cognitive processes that involve sequential transitions between thoughts—are to be found in a great number of domains of human cognition. These include mental simulation, planning, and any forms of multi-step inductive and deductive reasoning. Now notice that a program that provides accurate output even about a very complex domain can be very simple if it involves a search. The above-sketched hypothetical program that searches through all possible games of Go, or chess, and then assigns to a position its game-theoretic value by minimaxing is an example of such a program. Of course, such programs would take impractical amounts of time to deliver an output, and we would not want to call them 'intelligent'. Similar considerations apply to less extreme cases. An agent's reliance on processes that are more complex but take fewer steps to solve a problem (or to achieve a certain level of accuracy) will often count as more intelligent. A promising way to deal with this kind of problem that has been suggested elsewhere (Hernández-Orallo and Minaya-Collado 1998; Hernández-Orallo and Dowe 2010; Hernández-Orallo 2017) is to replace the notion of Kolmogorov complexity by that of Levin complexity (Levin 1973), which is defined as follows.[21]

$$\text{(Levin complexity)} \quad Kt(s) = min\,(l(p_s) + log\,S(p_s))$$

Here, $l(p_s)$ is the length of a program p that outputs the string s, and $S(p_s)$ is the number of steps taken by p to output s. If one bases one's notion of intelligence on Levin complexity, then systems score well just in case the algorithms they use represent a good trade-off between their length and the number of steps they require to provide an output. In many cases in which a system only uses a database or a value function, the number of steps taken is negligible. With respect to such cases, it is not important whether one measures intelligence by means of Kolmogorov complexity or by means of Levin complexity. But whenever some search-intensive algorithm is involved, Levin complexity appears to be more suitable for measuring intelligence. There is more to be said in favor of such an account. All cognitive systems face limitations both regarding their cognitive resources and the time they have for solving problems. Very fast processes/algorithms can be inadequate if they are too complex and thus consume too much storage space; the same applies to algorithms that are very short but take too much time. It seems natural to hold that intelligence requires finding the right trade-off in using these resources, which is captured by the notion of Levin complexity.

10 Conclusion

In this paper, I gave a new argument for the claim that intelligence requires data compression. The motivation for this view arose out of a paradox that seemed to show that either reliance on a lookup table manifests intelligence, or reliance on intuition does not: The paradox can be resolved if one accepts that intelligence depends on the complexity of the processes or algorithms used, and hence, on how well they are compressed. We saw that this is not the only reason to adopt such an account. Among other things, the relation between data compression and

[21] Legg and Hutter (2005, p. 435) also suggest that the notion of Levin complexity might be of use in an account of intelligence.

intelligence captures the intuition that making those, and only those, distinctions that matter is an essential part of intelligence. My discussion started from intuitive judgments in board games and their analogs in computer programs. However, as we saw, data compression is involved not only in processes applied to board games, and not only in intuitive judgments, but in large parts of cognition. This is no accident, since systems with limited cognitive resources need to compress data to be able to solve complex problems. As I argued, it as an important virtue of a theory of intelligence that it captures this fundamental empirical constraint. There are thus many reasons for acknowledging that intelligence requires data compression.

Acknowledgements I have presented versions of this article at the University of Rochester, the University of Cambridge, and the University of Osnabrück. I would like to thank the audiences on these occasions for helpful comments. I am especially grateful for comments and discussions to Scott Aaronson, José Hernández-Orallo, Joachim Horvath, Zeynep Soysal and two anonymous referees for this journal.

References

Aaronson, S. (2011). *Why philosophers should care about computational complexity*. arXiv:1108.1791.

Block, N. (1981). Psychologism and behaviorism. *Philosophical Review, 90,* 5–43.

Campbell, M., Hoane, J. H., Jr., & Hsu, F-h. (2002). Deep blue. *Artificial Intelligence, 134,* 57–83.

Chase, W., & Simon, H. (1973). The mind's eye in chess. In W. G. Chase (Ed.), *Visual information processing* (pp. 215–281). New York: Academic Press.

ChessOK. (2012). *Lomonosov tablebases*. Retrieved from http://tb7.chessok.com/. Accessed 5 Feb 2019.

Dowe, D., & Hajek, A. (1997). A computational extension to the turing test. *Technical Report #97/322*, Department of Computer Science, Monash University, Clayton 3168, Australia.

Dowe, D., Hernández-Orallo, J., & Das, P. (2011). Compression and intelligence: Social environments and communication. In J. Schmidhuber, K. R. Thorisson, & M. Looks (Eds.), *Artificial general intelligence. AGI 2011* (pp. 204–211). Berlin: Springer.

Dreyfus, H., & Dreyfus, S. (1986). *Mind over machine. The power of human intuition and expertise in the era of the computer*. New York: The Free Press.

Epstein, S. (2008). Intuition from the perspective of cognitive-experiential self-theory. In H. Plessner, C. Betsch, & T. Betsch (Eds.), *Intuition in judgment and decision making* (pp. 23–37). New York: Lawrence Erlbaum Associates.

Gigerenzer, G. (2007). *Gut feelings. The intelligence of the unconscious*. New York: Viking.

Gigerenzer, G., & Goldstein, D. (1996). Reasoning the fast and frugal way. Models of bounded rationality. *Psychological Review, 103*(4), 650–669.

Glöckner, A. (2008). Does intuition beat fast and frugal heuristics? a systematic empirical analysis. In H. Plessner, C. Betsch, & T. Betsch (Eds.), *Intuition in judgment and decision making* (pp. 309–325). New York: Lawrence Erlbaum Associates.

Glöckner, A., & Betsch, T. (2008). Multiple reason decision making based on automatic processing. *Journal of Experimental Psychology. Learning, Memory, and Cognition, 34*(5), 1055–1075.

Gobet, F., & Simon, H. (1996). Templates in chess memory: A mechanism for recalling several boards. *Cognitive Psychology, 31,* 1–40.

Hernández-Orallo, J. (2000). Beyond the turing test. *Journal of Logic, Language and Information, 9*(4), 447–466.

Hernández-Orallo, J. (2017). *The measure of all minds*. Cambridge: Cambridge University Press.

Hernández-Orallo, J., & Dowe, D. (2010). Measuring universal intelligence: Towards an anytime intelligence test. *Artificial Intelligence, 174*(18), 1508–1539.

Hernández-Orallo, J., & Minaya-Collado, N. (1998). A formal definition of intelligence based on an intensional variant of Kolmogorov complexity. In E. Alpaydin (Ed.), *Proceedings of the international ICSC symposium on engineering of intelligent systems* (pp. 146–163). Millet: ICSC Press.

Hutter, M. (2000). *A theory of universal artificial intelligence based on algorithmic complexity*. arXiv:cs.AI/0004001.

Hutter, M. (2005). *Universal artificial intelligence: Sequential decisions based on algorithmic probability*. Berlin: Springer.
Hutter, M. (2006). *The human knowledge compression prize*. http://prize.hutter1.net. Accessed 5 Feb 2019.
Kahneman, D. (2011). *Thinking, fast and slow*. New York: Farrar, Straus and Giroux.
Kahneman, D., & Klein, G. (2009). Conditions of intuitive expertise: A failure to disagree. *American Psychologist, 64*(6), 515–526.
Kasanoff, B. (2017, February 21). Intuition is the highest form of intelligence. *Forbes*. Retrieved from https://www.forbes.com/sites/brucekasanoff/2017/02/21/intuition-is-the-highest-form-of-intelligence/#4d0739583860. Accessed 5 Feb 2019.
Kasparov: 'Intuition Versus the Brute Force of Calculation'. (2003, February 10). *CNN.com*. Retrieved from http://edition.cnn.com/2003/TECH/fun.games/02/08/cnna.kasparov/. Accessed 5 Feb 2019.
Kasparov: 'Let Your Intuition Guide You'. (2017, March 13). *Goalcast.com*. Retrieved from https://goalcast.com/2017/03/13/chess-grandmaster-garry-kasparov-let-intuition-guide-you/. Accessed 5 Feb 2019.
Klein, G., Calderwood, R., & Clinton-Cirocco, A. (1985). Rapid decision making on the fire ground. In *Proceedings of the human factors and ergonomics society 30th annual meeting* (Vol. 1, pp. 576–580). Norwood, NJ: Ablex.
Kolmogorov, A. (1963). On tables of random numbers. *Sankhyā Series A, 25*, 369–375.
Legg, S., & Hutter, M. (2007). Universal intelligence: A definition of machine intelligence. *Minds and Machines, 17*, 391–444.
Levin, L. (1973). Universal sequential search problems. *Problems of Information Transmission, 9*(3), 265–266.
Li, M., & Vitányi, P. (2008). *An introduction to Kolmogorov complexity and its applications* (3rd ed.). New York: Springer.
Mahoney, M. (1999). Text compression as a test for artificial intelligence. In Proceedings of the National Conference on Artificial Intelligence, AAAI. Wiley, 970.
Myers, D. (2002). *Intuition. Its powers and perils*. New Haven: Yale University Press.
Okoli, J., Weller, G., & Watt, J. (2016). Information processing and intuitive decision-making on the fireground. Towards a model of expert intuition. *Cognition, Technology & Work, 18*(1), 89–103.
Pachur, T., & Marinello, G. (2013). Expert intuitions. How to model the decision strategies of airport customs officers? *Acta Psychologica, 144*, 97–103.
Poland, J., & Hutter, M. (2005). Asymptotics of discrete MDL for online prediction. *IEEE Transactions on Information Theory, 51*, 3780–3795.
Poland, J., & Hutter, M. (2006). MDL convergence speed for Bernoulli sequences. *Statistics and Computing, 16*(2), 161–175.
Shannon, C., & McCarthy, J. (1956). Preface. In ibid. *Automata Studies, 34*(v), viii.
Silver, D., et al. (2017). Mastering the game of go without human knowledge. *Nature*,. https://doi.org/10.1038/nature24270.
Silver, D., et al. (2018). A general reinforcement learning algorithm that masters chess, shogi, and go through self-play. *Science, 362*(6419), 1140–1144.
Simon, H. (1995). Explaining the ineffable AI on the topics of intuition, insight and inspiration. In C. Mellish (Ed.), *IJCAI'95 Proceedings of the 14th international joint conference on artificial intelligence* (Vol. 1, pp. 939–948)., Morgan Kaufmann Publishers CA: San Francisco.
Solomonoff, R. (1964). A formal theory of inductive inference: Parts 1 and 2. *Information and Control, 7*, 1–22.
Stockfish Evaluation Guide: Piece Value mg. (2018, January). Retrieved from https://hxim.github.io/Stockfish-Evaluation-Guide/. Accessed 5 Feb 2019.
Turing, A. (1953). Chess. In B. V. Bowden (Ed.), *Faster than thought*. London: Pitman & Sons.
Von Neumann, J. (1928). Zur Theorie der Gesellschaftsspiele. *Mathematische Annalen, 100*(1), 295–320.
Weinberg, J., et al. (2010). Are philosophers expert intuiters? *Philosophical Psychology, 23*(3), 331–355.

Synthese (2021) 198 (Suppl 27):S6491–S6504
https://doi.org/10.1007/s11229-019-02148-2

S.I.: DECTHEORY&FUTOFAI

Approval-directed agency and the decision theory of Newcomb-like problems

Caspar Oesterheld[1,2]

Received: 17 January 2018 / Accepted: 15 February 2019 / Published online: 25 February 2019

Abstract

Decision theorists disagree about how instrumentally rational agents, i.e., agents trying to achieve some goal, should behave in so-called Newcomb-like problems, with the main contenders being causal and evidential decision theory. Since the main goal of artificial intelligence research is to create machines that make instrumentally rational decisions, the disagreement pertains to this field. In addition to the more philosophical question of what the right decision theory is, the goal of AI poses the question of how to implement any given decision theory in an AI. For example, how would one go about building an AI whose behavior matches evidential decision theory's recommendations? Conversely, we can ask which decision theories (if any) describe the behavior of any existing AI design. In this paper, we study what decision theory an approval-directed agent, i.e., an agent whose goal it is to maximize the score it receives from an overseer, implements. If we assume that the overseer rewards the agent based on the expected value of some von Neumann–Morgenstern utility function, then such an approval-directed agent is guided by two decision theories: the one used by the agent to decide which action to choose in order to maximize the reward and the one used by the overseer to compute the expected utility of a chosen action. We show which of these two decision theories describes the agent's behavior in which situations.

Keywords Reinforcement learning · Causal decision theory · Evidential decision theory · Newcomb's problem · AI safety · Philosophical foundations of AI

Chapter 5 was originally published as Oesterheld, C. Synthese (2021) 198 (Suppl 27): S6491–S6504. https://doi.org/10.1007/s11229-019-02148-2.

✉ Caspar Oesterheld
caspar.oesterheld@foundational-research.org; caspar.oesterheld@duke.edu

1 Foundational Research Institute, Berlin, Germany

2 Present Address: Duke University, Durham, USA

1 Introduction

In decision theory, there is a large debate about how an instrumentally rational agent, i.e., an agent trying to achieve some goal or maximize some utility function, should decide in Newcomb's problem (introduced by Nozick 1969) and variations thereof (a list is given by Ledwig 2000, pp. 80–87). Consequently, different normative theories of instrumental rationality have been developed. The best known ones are evidential (sometimes also called Bayesian) decision theory (EDT) (Ahmed 2014; Almond 2010; Price 1986; Horgan 1981) and causal decision theory (CDT) (Gibbard and Harper 1981; Joyce 1999; Lewis 1981; Skyrms 1982; Weirich 2016), but many have attempted to remediate what they view as failures of the two theories by proposing further alternatives (Spohn 2003, 2012; Poellinger 2013; Arntzenius 2008; Gustafsson 2011; Wedgwood 2013; Dohrn 2015; Price 2012; Soares and Levinstein 2017).

Because the main goal of artificial intelligence is to build machines that make instrumentally rational decisions (Russell and Norvig 2010, Sects. 1.1.4, 2.2; Legg and Hutter 2007; Doyle 1992), this normative disagreement has some bearing on how to build these machines (cf. Soares and Fallenstein 2014a, Sect. 2.2; Soares and Fallenstein 2014b, Sect. 1; Bostrom 2014b, Chap. 13, Sect. "Decision theory"). The differences between these decision theories are probably inconsequential in most situations (Ahmed 2014, Sect. 0.5, Chap. 4; Briggs 2017),[1] but still matter in some (Ahmed 2014, Chap. 4–6; Soares 2014a; Bostrom 2014a). In fact, AI may expose the differences more often. For example, Newcomb's problem and the prisoner's dilemma with a replica (Kuhn 2017, Sect. 7) are easy to implement for agents with copyable source code (cf. Yudkowsky 2010 pp. 85ff. Soares and Fallenstein 2014b, Sect. 2; Soares 2014b; Cavalcanti 2010; Sect. 5). Indeed, the existence of many copies is the norm for (successful) software, including AI-based software. While copies of present-day software systems may only interact with each other in rigid, explicitly pre-programmed ways, future AI-based systems will make decisions in a more autonomous, flexible and goal-driven way. Overall, the decision theory of Newcomb-like scenarios is a central foundational issue which will plausibly become practically important in the longer term.

The problem for AI research posed by the disagreement among decision theorists can be divided into two questions:

1. What decision theory do we want an AI to follow?
2. How could we implement such a decision theory in an AI? Or: How do decision theories and AI frameworks or architectures map onto each other?

Although it certainly requires further discussion, there already is a large literature related to the first question.[2] In this paper, I would thus like to draw attention to the second question.

[1] In fact, Eells (1981) has argued that EDT and CDT always behave in the same way, but I disagree with this assessment based on the reasons given by Ahmed (2014, Sect. 4.3–4.6) and Price (1986).

[2] Of course, the existing literature asks about the right decision theory proper. The answer to that question might differ from the answer to the AI-specific question (cf. Kumar 2017; Treutlein 2018). After all, even if we have identified the right decision theory for ourselves, we may want to implement a different decision theory in an AI. One reason could be that the main contenders are not self-recommending—it has been pointed out that EDT and CDT both recommend to self-modify into slightly different decision theories

Specifically, I would like to investigate how approval-directed agents behave in Newcomb-like problems. By an approval-directed agent, I mean an agent that is coupled with an overseer. After the agent has chosen an action, the overseer scores the agent for that action. Rather than, say, trying to bring about particular states in the environment, the agent chooses actions so as to maximize the score it receives from the overseer (cf. Christiano 2014). A model of approval-directed agency that allows us to describe Newcomb-like situations is described and discussed in Sect. 2.

Approval-directed agency is intended as a model of reinforcement learning agents (see Sutton and Barto 1998; Russell and Norvig 2010; Chaps. 17, 21, for introductions to reinforcement learning), for whom the reward function is analogous to the approval-directed agent's overseer. Since reinforcement learning is such a general and commonly studied problem in artificial intelligence (Hutter 2005, e.g. Chap. 4.1.3; Russell and Norvig 2010, p. 831; Sutton and Barto 1998, Chap. 1), it is an especially attractive target for modeling.[3] However, because decision theories are usually defined only for single decisions, we will only discuss single decisions whereas reinforcement learning is usually concerned with sequential interactions of agent and environment. However, this decision can also be a policy choice to model sequential decision problems.[4] In addition to limiting our analysis to single decisions, we will not discuss the learning process and simply assume that the agent has already formed some model of the world.

If we assume that, after an action has been taken, the overseer rewards the agent based on the expected value of some von Neumann–Morgenstern utility function, the agent is implicitly driven by two decision theories: The overseer can use the regular conditional expectation or the causal expectation to estimate the value of its utility function; and the agent itself can follow CDT or EDT when maximizing the score it receives from the overseer (Sect. 3).

We then show how the overall decision theory depends on these two potentially conflicting decision theories. If the overseer bases its expected value calculations on looking only at the world, then the agent's decision theory is decisive. If the overseer

Footnote 2 continued
(Meacham 2010; Soares and Fallenstein 2014b, Sect. 3; Yudkowsky 2010, Sect. 2; Greene 2018). The same arguments imply that even if one is convinced of CDT or EDT one would not want the AI to use CDT and EDT. That said, one could also leave the self-modification to the AI.

3 Reinforcement learning and approval-directed agency are also common outside of artificial intelligence. For example, Achen and Bartels (2016, Chap. 4) review evidence which shows that electorates often vote retrospectively to punish or reward incumbents.

4 This is consistent with what reinforcement learning algorithms usually do—they choose policies rather than individual actions. This is because the utility of a single action usually cannot be evaluated without knowing how the agent will deal with situations that might arise as a result of taking that action.
When individual actions *can* be evaluated in isolation, the *ex ante* policy choice sometimes differs from the choice of individual actions (see the absent-minded driver, introduced by Piccione and Rubinstein 1997; cf. Aumann et al. 1997; the Newcomb-like scenarios discussed by, e.g., Hintze 2014; Soares and Levinstein 2017, Sect. 2; and the problems in anthropics discussed by Armstrong 2011). While it is rarely discussed in the debate between evidential and causal decision theorists, a few authors regard this discrepancy as crucial and have argued that a proper decision theory should be about optimal policy choices (e.g. Hintze 2014; Soares and Fallenstein 2014b, Sect. 2.1; Soares and Levinstein 2017, Sect. 2). However, this issue is beyond the scope of the present paper.
Further issues in sequential Newcomb-like problems are discussed by Everitt et al. (2015).

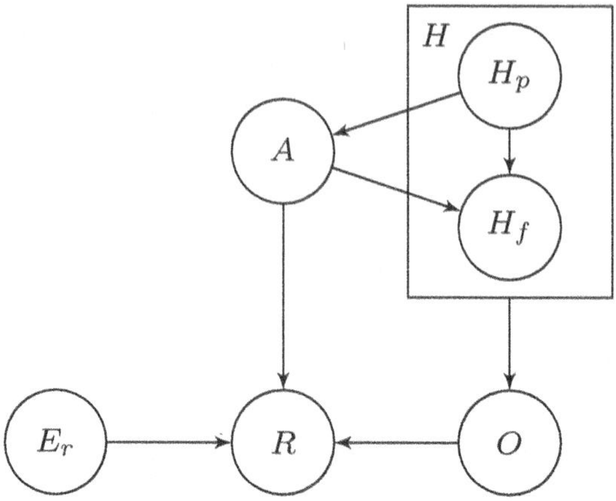

Fig. 1 A causal model of an approval-directed agent in a Newcomb-like decision problem. A denotes the agent's action, H the environment history, O the observation on which the overseer bases the reward, R is that reward, and E_r is information about the way the reward is computed that is only available to the overseer. The box is used to indicate that H includes the two random variables H_p and H_f. All of H may have a causal influence on O

bases its estimates only on the agent's action, then the overseer's decision (or perhaps rather action evaluation) theory is decisive.

2 Approval-directed agency

We first describe a model of approval-directed agency. To be able to apply both CDT and EDT, we will use causal models in Pearl's (2009) sense. Consequently, we use Pearl's *do*-calculus-based version of CDT (Pearl 2009, Chap. 4). We will, throughout this paper, assume that the agent has already formed a (potentially implicit) model of the world[5]—e.g., based on past interactions with the environment. Also, we will only consider single decisions rather than sequential problems of iterative interaction between agent and environment.

A causal model of such a one-shot Newcomb problem from the perspective of the approval-directed agent is given in Fig. 1. In this model, the agent decides to take some action A, which may causally affect some part of the environment history, i.e., the history of states, H. We will call that part of the history the agent's causal future H_f. Furthermore, the agent may be causally influenced by some other part of the environment history, which we will call the agent's causal past H_p. H may contain information other than H_f and H_p, which we will assume to be independent of A.[6] The *overseer*, physically realized by, e.g., some module physically attached to the agent or

[5] There is a broad philosophical literature on whether causal relationships exist and whether they can be inferred in cases where the agent is part of the environment. See, e.g., the edited volume by Price and Corry (2007).

[6] For simplicity, we will ignore dependences not resulting from causation (Arntzenius 2010). For example, if you play against a copy, there is a logical dependence between your and your copy's decision. Even if you know a set of nodes in the causal graph that d-separates your and your copy's decision (e.g., if you

a human supervisor, observes the agent's action and partially, via some percept O, the state of the world[7]. The overseer then calculates the reward R. To set proper incentives to the agent, we will assume the overseer to know not only the action and observation, but also everything that the agent knows (cf. Christiano 2016). The overseer may also have access to some additional piece of information E_r about the way the reward is to be calculated.[8] Lastly, we assume that the sets of possible values of A, O and E_r are finite.

In principle, the overseer could reward the agent in all kinds of ways. E.g., it could reward the agent "deontologically" (Alexander and Moore 2016) for taking a particular action independently of the consequences of taking that action. In this paper, we will assume that the reward estimates the value of some von Neumann–Morgenstern utility function U that only depends on states of the world. I use the capital U to indicate that the utility function, too, is a random variable (in the Bayesian sense). For simplicity's sake, we will, again, assume that the set of possible values of U is finite.

We will view U as representing the system designer's preferences over world states.[9] While other ways of assigning the reward are possible, this is certainly an attractive way of getting an approval-directed agent to achieve goals that we want it to achieve. After all, in real-world applications, we will usually care about the outcomes of the agent's decisions, such as whether a car has reached its destination in time or whether a human has been hurt.

The standard way of estimating $U(H)$ (or any quantity for that matter) is the familiar conditional expectation. Thus, the overseer may compute the reward as

$$r = \mathbb{E}\left[U(H) \mid e_r, a, o\right], \tag{1}$$

Footnote 6 continued
know your common source code), the dependence persists. We exclude these dependences because such situations cannot be modeled by standard causal graphs.
However, we could adapt causal graphs to accomodate for these kinds of dependences. First, we could modify our definition of causality in such a way that dependence does imply causation, as has been proposed by Spohn (2003, 2012), Yudkowsky (2010) and others. For instance, we could model the dependence between the outputs of two instances of an algorithm by introducing a logical node as a common cause of the two. This logical node would then represent the output of the abstract algorithm that the two copies implement. While changes to the concept of causation may affect CDT's implied behavior, the results from this paper can be directly transfered to such modifications.
Alternatively, we could extend causal graphs to also include non-causal dependences (cf. Poellinger 2013). Such extension necessitates a new CDT formalism, so the proofs from this paper do not directly transfer to this case. That said, I expect our results to generalize given that both EDT and CDT would probably treat non-causal dependences on the action just like they treat causal arrows directed toward the action.

7 Christiano (2014) does not define approval-directed agency formally, but judging from a comment he made at https://medium.com/paulfchristiano/i-agree-that-the-key-feature-of-approval-directed-agents-is-that-the-causal-picture-is-736b4474910e, he considers it crucial to his conception that the overseer only looks at the agent's action and does not observe the action's consequences (cf. the distinction introduced in Sect. 3).

8 One reason for the overseer to have access to such additional information is that some of the human supervisor's values may not be expressible in a way that the approval-directed agent's algorithm can utilize (cf. Muehlhauser and Helm 2012, Sects. 3, 4, 5.3).

9 Some have tried to modify the reward relative to the designer's preferences to make the reinforcement learning problem easier to solve (Sorg 2011), although Sutton and Barto (1998, Sect. 3.2) explicitly discourage such tricks in their reinforcement learning textbook.

where r, a, e_r, and o are values of R, A, E_r, and O, respectively.[10]

A causal decision theorist overseer agrees that after an action a is taken the right-hand side of Eq. 1 most accurately estimates how much utility is achieved. She merely thinks that this term should not be used to decide which action a to take in the first place.[11] However, this puts a causal decision theorist overseer in a peculiar situation. Whatever formula she uses to compute the reward will also be used by the reward-maximizing agent to decide which action to take. A causal decision theorist overseer might therefore worry (rightfully, as we will see) that providing rewards according to Eq. 1 will make the agent EDT-ish. Hence, she either has to incorrectly estimate how much utility was achieved; or live with the agent using an—in her mind—incorrect way of weighing her options. If she prefers the latter, she would reward according to Eq. 1. But arguably getting the agent to choose correctly is the overseer's primary goal. Thus, she might prefer to compute the reward according to

$$r = \mathbb{E}\left[U(H) \mid e_r, do(a), o\right]. \tag{2}$$

Here, $do(a)$ refers to Pearl's do-calculus, where conditioning on $do(a)$ roughly means intervening from outside the causal model to set A to a. For an introduction to the do-calculus, see Pearl (2009). Although a causal decision theorist overseer may prefer computing rewards according to Eq. 1, we will from now on say "the overseer uses CDT" if rewards are computed according to Eq. 2 and "the overseer uses EDT" if rewards are calculated according to Eq. 1.

An approval-directed agent is characterized by maximizing the reward it receives from the overseer.[12] However, decision theory offers us, again, (at least) two different expected values, the regular expected value of EDT

$$\mathbb{E}\left[R \mid a\right], \tag{3}$$

and CDT's causal expected value

$$\mathbb{E}\left[R \mid do(a)\right]. \tag{4}$$

[10] At first sight this may be confusing to some readers, because in reinforcement learning, utility sometimes refers to expected cumulative reward (Russell and Norvig 2010, Chap. 17, 21), although others use the term *value function* instead (Sutton and Barto 1998, Sect. 3.7). Here, U does not refer to utility in that sense but in the decision-theoretical sense of representing intrinsic values. So, in the present case, we have two "layers" of goals: first, the agent maximizes the reward r. Second, the agent as incentivized by the overseer's way of calculating rewards maximizes utility $U(H)$.
One cause of confusion is that in model applications of reinforcement learning, the reward function possesses full knowledge of the world state and thus does not require the use of the expectation operator.

[11] If the disagreement in Newcomb's problem is to be about different theories of rational choice (EDT, CDT and so forth) rather than the predictive abilities of "the being", Omega or the psychologist, then after requesting both boxes a proponent of two-boxing has to believe that she will probably receive only \$1000. Causal and evidential decision theorists agree that regular conditional expectation is the correct way of updating one's beliefs about the state of the world after an action has been taken (cf. the distinction between "acts" and "actions" in Pearl 2009 Sect. 4.1.1).

[12] In reinforcement learning, some have proposed alternative optimization targets that incorporate, e.g., risk aversion (García and Fernández 2015, Sect. 3).

We leave the interesting question of which (if any) decision theory describes the behavior of current reinforcement learning algorithms to future research[13] and in the following assume that the agent is known to implement either CDT or EDT.

3 The conflict of the decision theories of agent and overseer

When viewed together with the overseer, our agent may now be seen as containing two decision theories, one for computing the reward and one in the algorithm that tries to find the action to maximize that reward. These decision theories may not always be the same. Given this potential discrepancy, the question is which of the two decision theories prevails, i.e., for which configurations of the two decision theories the overall agent acts like a CDT agent and for which it acts like an EDT agent w.r.t. U.

As it turns out, the answer to this question depends on the decision problem in question. In particular, it depends on whether the overseer updates its estimate of $U(H)$ primarily based on the action taken by the agent or on its observation of the environment.

For illustration, consider two versions of Newcomb's problem. In both versions, the predictor is equally reliable—e.g., correct with 90% probability—and the potential box contents are the same—e.g., the standard \$1K and \$1M. As usual, the content of the opaque box cannot be causally influenced by one's decision. In the first version, the overseer eventually sees the payoff, i.e., how much money the agent has made. In this case, as soon as the money is observed, the overseer's estimate of $U(H)$ becomes independent of the agent's action. More generally, O may tell the overseer so much about $U(H)$ that it becomes independent of A even if $U(H)$ is not yet fully observed. That is,

$$\mathbb{E}[U(H) \mid e_r, a, o] = \mathbb{E}[U(H) \mid e_r, o] \tag{5}$$

and

$$\mathbb{E}[U(H) \mid e_r, do(a), o] = \mathbb{E}[U(H) \mid e_r, o] \tag{6}$$

for all e_r, a and o. Note that neither of these two implies the other.[14] Intuitively speaking, these two mean that the reward is ultimately determined by $U(H)$.

In the second version of Newcomb's problem, the monetary payoff is not observed but covertly invested into increasing the agent's utility function. Only the agent's choice can then inform the overseer about $U(H)$. Formally, it is both

[13] For preliminary work on this question, see Mayer et al. (2016), Oesterheld (2018a) and perhaps Albert and Heiner (2001).

[14] We give a brief justification of this claim. If all of a's causal influence on H can be discerned from O, then, of course, a could still be diagnostically relevant for one's estimate of $U(H)$. The other direction is more complicated. The idea is that Eq. 5 can be true if the causal and non-causal implications of a exactly cancel each other out. An example is a version of Newcomb's problem in which one-boxing ensures with certainty that both boxes contain the same amount of money. Then if O and E_r do not contain any information, the expected value of two-boxing and one-boxing is the same and so learning of the action is irrelevant for estimating $U(H)$. However, two-boxing is causally better than one-boxing, so Eq. 6 is violated.

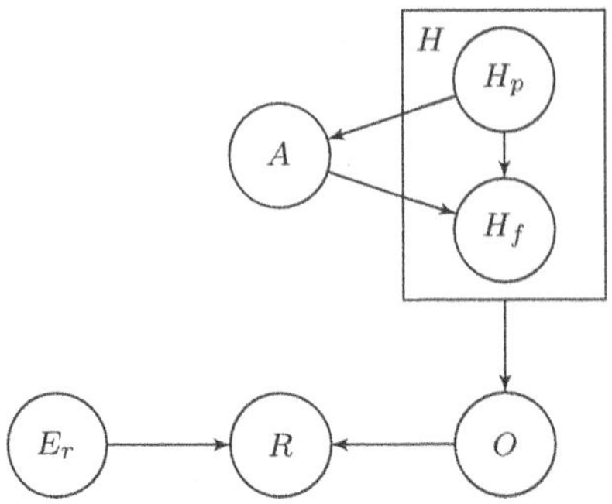

(a) The reward is computed based on observing the state of the world.

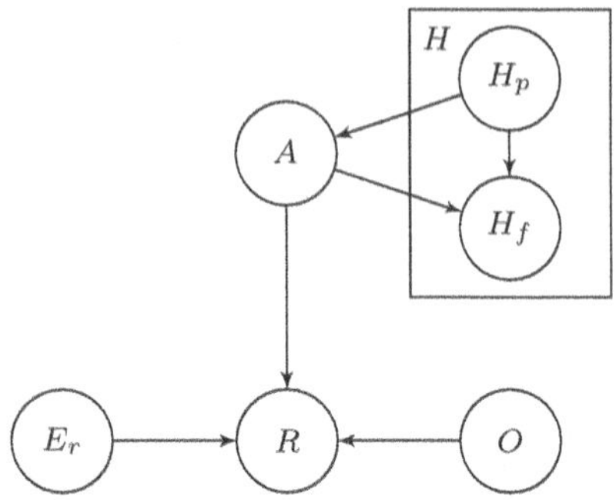

(b) The reward is computed based on observing the agent's action.

Fig. 2 Two different ways in which the overseer can calculate the reward

$$\mathbb{E}\left[U(H) \mid e_r, a, o\right] = \mathbb{E}\left[U(H) \mid e_r, a\right] \tag{7}$$

and

$$\mathbb{E}\left[U(H) \mid e_r, do(a), o\right] = \mathbb{E}\left[U(H) \mid e_r, do(a)\right]. \tag{8}$$

Intuitively speaking, these two equations mean that the reward is not determined by $U(H)$ but by what the overseer believes $U(H)$ will be given a or $do(a)$.

Again, we assume that this is known to the agent. An example class of cases is that in which the agent's decisions are correlated with those of agents in far-away parts of the environment (cf. Treutlein and Oesterheld 2017; Oesterheld 2018b). The two versions are depicted in Fig. 2.

Of course, these are only the two extremes from the set of all possible situations. In real-world Newcomb-like scenarios, the overseer may also draw some information from both sources. Nonetheless, it seems useful to understand the extreme cases, as this may also help us understand mixed ones.

In the following subsections, we will show that in the first type, the decision theory of the agent is decisive, whereas in the second type, the overseer's decision theory is[15]. Roughly, the reason for that is the following: As noted earlier, the reward in the first type depends directly on $U(H)$. Thus, the agent will try to maximize $U(H)$ according to its own decision theory. In the second type, the overseer takes the agent's action a and then considers what either a or $do(a)$ says about $U(H)$. Thus, the agent has to pay careful attention to whether the overseer uses EDT's or CDT's expected value.

We prove this formally by considering all possible configurations of the type of the problem, the overseer's decision theory and the agent's decision theory. While we will limit our analysis to EDT and CDT, the results can easily be generalized to variants of these that arise from modifying the causal model or conditional credence distribution (e.g. Yudkowsky 2010; "Disposition-based decision theory"; Spohn 2012; Dohrn 2015). The analysis is summarized in Table 1.

[15] The dominance of the overseer's decision theory in the second type of Newcomb's problem is mentioned (though not proven) by Christiano (2014, Sect. "Avoid lock-in").

Table 1 An overview of the results of the calculations in Sect. 3

Type of Newcomb problem	Agent's DT	Overseer's DT	Resulting DT
First type	CDT	EDT	CDT
		CDT	CDT
	EDT	EDT	EDT
		CDT	EDT
Second type	CDT	EDT	EDT
		CDT	CDT
	EDT	EDT	EDT
		CDT	CDT

3.1 First type

3.1.1 The EDT agent

The EDT agent judges its action by

$$\mathbb{E}[R \mid a]. \tag{9}$$

If the overseer calculates regular conditional expectation, then it is

$$\mathbb{E}[R \mid a] = \mathbb{E}[\mathbb{E}[U(H) \mid E_r, O, a] \mid a] \tag{10}$$
$$= \mathbb{E}[U(H) \mid a], \tag{11}$$

where the last line is due to what is sometimes called the law of total expectation (LTE) or the tower rule (see, e.g., Ross 2007, Sect. 3.4; Billingsley 1995, Theorem 34.4). Intuitively, you cannot expect that gaining more evidence (i.e., E_r and O in addition to a) moves your expectation of $U(H)$ into any particular direction.

Because the overseer knows more than the agent, we will need this rule in all of the following derivations. Its application makes it hard to generalize these results to other decision theories, since LTE does not apply if the two decision theories do not both compute a form of expected utility.

Equations 10 and 11 show that if the overseer computes regular expected value and the agent maximizes the reward according to EDT, then the agent as a whole maximizes U according to EDT.

If the overseer computes CDT's expected value, it is

$$\mathbb{E}[R \mid a] = \mathbb{E}[\mathbb{E}[U(H) \mid E_r, do(a), O] \mid a] \tag{12}$$
$$= \sum_{e_r, o} P(e_r, o \mid a) \cdot \mathbb{E}[U(H) \mid e_r, do(a), o] \tag{13}$$
$$\underset{\text{eq. 5 and 6}}{=} \sum_{e_r, o} P(e_r, o \mid a) \cdot \mathbb{E}[U(H) \mid e_r, a, o] \tag{14}$$

$$= \mathbb{E}\left[\mathbb{E}\left[U(H) \mid E_r, a, O\right] \mid a\right] \tag{15}$$

$$\underset{\text{LTE}}{=} \mathbb{E}\left[U(H) \mid a\right] \tag{16}$$

3.1.2 The CDT agent

The CDT agent judges its action by

$$\mathbb{E}\left[R \mid do(a)\right]. \tag{17}$$

If the overseer uses regular expected value (EDT), then

$$\begin{aligned}
\mathbb{E}\left[R \mid do(a)\right] &= \mathbb{E}\left[\mathbb{E}\left[U(H) \mid a, O, E_r\right] \mid do(a)\right] & (18)\\
&= \sum_{e_r, o} P(e_r, o \mid do(a)) \cdot \mathbb{E}\left[U(H) \mid a, o, e_r\right] & (19)\\
&\underset{\text{eq. 5 and 6}}{=} \sum_{e_r, o} P(e_r, o \mid do(a)) \cdot \mathbb{E}\left[U(H) \mid do(a), o, e_r\right] & (20)\\
&= \mathbb{E}\left[\mathbb{E}\left[U(H) \mid do(a), O, E_r\right] \mid do(a)\right] & (21)\\
&\underset{\text{LTE}}{=} \mathbb{E}\left[U(H) \mid do(a)\right] & (22)
\end{aligned}$$

Learning about an intervention $do(a)$ cannot always be treated in the same way as learning about other events. Hence, the application of the law of total expectation is not straightforward. However, $P(\cdot \mid do(x))$ is always a probability distribution. Because the law of total expectation applies to all probability distributions, it also applies to ones resulting from the application of the do-calculus.

If the overseer uses CDT's expected value, then

$$\begin{aligned}
\mathbb{E}\left[R \mid do(a)\right] &= \mathbb{E}\left[\mathbb{E}\left[U(H) \mid E_r, O, do(a)\right] \mid do(a)\right] & (23)\\
&\underset{\text{LTE}}{=} \mathbb{E}\left[U(H) \mid do(a)\right]. & (24)
\end{aligned}$$

3.2 Second type

3.2.1 The EDT agent

The EDT agent judges its actions by

$$\mathbb{E}\left[R \mid a\right]. \tag{25}$$

If the overseer is based on regular conditional expectation (EDT), then it is again

$$\begin{aligned}
\mathbb{E}\left[R \mid a\right] &= \mathbb{E}\left[\mathbb{E}\left[U(H) \mid E_r, a\right] \mid a\right] & (26)\\
&\underset{\text{LTE}}{=} \mathbb{E}\left[U(H) \mid a\right]. & (27)
\end{aligned}$$

If the overseer is based on CDT-type expectation, then

$$\mathbb{E}[R \mid a] = \mathbb{E}[\mathbb{E}[U(H) \mid E_r, do(a)] \mid a] \tag{28}$$
$$= \sum_{e_r} P(e_r \mid a) \cdot \mathbb{E}[U(H) \mid do(a), e_r] \tag{29}$$
$$= \sum_{e_r} P(e_r) \cdot \mathbb{E}[U(H) \mid do(a), e_r] \tag{30}$$
$$= \sum_{e_r} P(e_r \mid do(a)) \cdot \mathbb{E}[U(H) \mid do(a), e_r] \tag{31}$$
$$= \mathbb{E}[\mathbb{E}[U(H) \mid E_r, do(a)] \mid do(a)] \tag{32}$$
$$\underset{\text{LTE}}{=} \mathbb{E}[U(H) \mid do(a)]. \tag{33}$$

3.2.2 The CDT agent

The CDT agent judges actions by

$$\mathbb{E}[R \mid do(a)]. \tag{34}$$

Because of Rule 2 in Theorem 3.4.1 of Pearl (2009, Sect. 3.4.2) applied to the causal graph of Fig. 2b, it is

$$\mathbb{E}[R \mid do(a)] = \mathbb{E}[R \mid a]. \tag{35}$$

Thus, the analysis of the CDT agent is equivalent to that of the EDT agent.

4 Conclusion

In this paper, we have taken a step to map reinforcement learning architectures onto decision theories. We found that in Newcomb-like problems, if the overseer rewards the agent purely on the basis of the agent's action, then the overall system's behavior is determined by the decision theory implicit in the overseer's reward function. If the overseer judges the agent based on looking at the world, however, then the agent's decision theory is decisive.

This has implications for how we should design approval-directed agents. For instance, if we would like to leave decision-theoretical judgements to the overseer, we must ensure that the overseer assigns rewards before making new observations about the world state (cf. Christiano 2014, Sect. "Avoid lock-in"). Of course, this makes the reward less accurate and may thus slow down the agent's learning process. If we want the overseer to look at both the world and the agent's action, then we need to align both the overseer's and the agent's decision theory.

Much more research is left to be done at the intersection of decision theory and artificial intelligence. For instance, what (if any) decision theories describe the way modern reinforcement learning algorithms maximize reward? Do the results of this paper generalize to sequential decision problems? Moving away from the reinforcement learning framework, what decision theories do other frameworks in AI implement? What about decision theories other than CDT and EDT?

Acknowledgements I am indebted to Max Daniel, Johannes Treutlein, Tom Everitt, Lukas Gloor, Sören Mindermann, Brian Tomasik and Tobias Baumann for valuable comments and discussions.

References

Achen, C. H., & Bartels, L. M. (2016). *Democracy For realists. Why elections do not produce responsive government*. Princeton: Princeton University Press.

Ahmed, A. (2014). *Evidence, decision and causality*. Cambridge: Cambridge University Press.

Albert, M., & Heiner, R. A. (2001). *An indirect-evolution approach to Newcomb's problem*. CSLE discussion paper, no. 2001-01. https://www.econstor.eu/bitstream/10419/23110/1/2001-01_newc.pdf. Accessed 22 Feb 2019.

Alexander, L., & Moore, M. (2016). Deontological ethics. In E. N. Zalta (Ed.), *The Stanford encyclopedia of philosophy*. Winter 2016. Metaphysics Research Lab, Stanford University. https://plato.stanford.edu/archives/win2016/entries/ethics-deontological/. Accessed 22 Feb 2019.

Almond, P. (2010). *On causation and correlation part 1: Evidential decision theory is correct*. https://casparoesterheld.files.wordpress.com/2016/12/almond_edt_1.pdf. Accessed 22 Feb 2019.

Armstrong, S. (2011). *Anthropic decision theory*. Future of Humanity Institute. arXiv: 1110.6437.

Arntzenius, F. (2008). No regrets, or: Edith Piaf revamps decision theory. *Erkenntnis*, *68*(2), 277–297. https://doi.org/10.1007/s10670-007-9084-8.

Arntzenius, F. (2010). Reichenbach's common cause principle. In E. N. Zalta (Ed.), *The Stanford encyclopedia of philosophy*. Fall 2010. Metaphysics Research Lab, Stanford University.

Aumann, R. J., Hart, S., & Perry, M. (1997). The absent-minded driver. *Games and Economic Behavior*, *20*, 102–116.

Billingsley, P. (1995). *Probability and measure* (3rd ed.). Hoboken: Wiley.

Bostrom, N. (2014a). *Hail mary, value porosity, and utility diversification*. http://www.nickbostrom.com/papers/porosity.pdf. Accessed 22 Feb 2019.

Bostrom, N. (2014b). *Superintelligence. Paths, dangers, strategies* (1st ed.). Oxford: Oxford University Press.

Briggs, R. (2017). Real-life Newcomb problems? In *Talk at the 1st workshop on decision theory & the future of artificial intelligence in Cambridge, UK*.

Cavalcanti, E. G. (2010). Causation, decision theory, and Bell's theorem: A quantum analogue of the Newcomb problem. *The British Journal for the Philosophy of Science*, *61*(3), 569–597. https://doi.org/10.1093/bjps/axp050.

Christiano, P. (2014). *Model-free decisions*. https://ai-alignment.com/model-free-decisions-6e6609f5d99e. Accessed 22 Feb 2019.

Christiano, P. (2016). *Adequate oversight*. https://ai-alignment.com/adequate-oversight-25fadf1edce9. Accessed 22 Feb 2019.

Dohrn, D. (2015). Egan and agents: How evidential decision theory can deal with Egan's dilemma. *Synthese*, *192*(6), 1883–1908.

Doyle, J. (1992). Rationality and its roles in reasoning. *Computational Intelligence*, *8*(2), 376–409.

Eells, E. (1981). Causality, utility, and decision. *Synthese*, *48*(2), 295–329.

Everitt, T., Leike, J., & Hutter, M. (2015). Sequential extensions of causal and evidential decision theory. In T. Walsh (Ed.), *Algorithmic decision theory: 4th international conference, ADT 2015, Lexington, KY, USA, September 27–30, 2015, Proceedings* (pp. 205–221). Springer. https://doi.org/10.1007/978-3-319-23114-3_13.

Fisher, J. C. *Disposition-based decision theory*. https://casparoesterheld.files.wordpress.com/2019/02/dbdt.pdf. Accessed 22 Feb 2019.

García, J., & Fernández, F. (2015). A comprehensive survey on safe reinforcement learning. *Journal of Machine Learning Research*, *16*, 1437–1480.
Gibbard, A., & Harper, W. L. (1981). Counterfactuals and two kinds of expected utility. In W. L. Harper, R. Stalnaker, & G. Pearce (Eds.), *Ifs. Conditionals, belief, decision, chance and time* (Vol. 15). The University of Western Ontario Series in Philosophy of Science. A series of books in philosophy of science, methodology, epistemology, logic, history of science, and related fields (pp. 153–190). Springer. https://doi.org/10.1007/978-94-009-9117-0_8.
Greene, P. (2018). Success-first decision theories. In A. Ahmed (Ed.), *Newcomb's problem*. Classic Philosophical Arguments. Cambridge University Press. https://doi.org/10.1017/9781316847893.007.
Gustafsson, J. E. (2011). A note in defence of ratificationism. *Erkenntnis*, *75*(1), 147–150.
Hintze, D. (2014). *Problem class dominance in predictive dilemmas*. http://intelligence.org/files/ProblemClassDominance.pdf. Accessed 22 Feb 2019.
Horgan, T. (1981). Counterfactuals and Newcomb's problem. *The Journal of Philosophy*, *78*(6), 331–356.
Hutter, M. (2005). *Universal artificial intelligence. sequential decision based on algorithmic probability*. In W. Brauer, G. Rozen-berg, & A. Salomaa (Eds.), Texts in theoretical computer science. Springer.
Joyce, J. M. (1999). *The foundations of causal decision theory. Cambridge studies in probability, induction, and decision theory*. Cambridge: Cambridge University Press.
Kuhn, S. (2017). Prisoner's dilemma. In E. N. Zalta (Ed.), *The Stanford encyclopedia of philosophy*. Spring 2017. Metaphysics Research Lab, Stanford University. https://plato.stanford.edu/archives/spr2017/entries/prisoner-dilemma/. Accessed 22 Feb 2019.
Kumar, R. (2017). New work for decision theorists. In *Talk at the 1st workshop on decision theory & the future of artificial intelligence in Cambridge, UK*.
Ledwig, M. (2000). *Newcomb's problem*. Ph.D. thesis, University of Constance. https://kops.uni-konstanz.de/bitstream/handle/123456789/3451/ledwig.pdf. Accessed 22 Feb 2019.
Legg, S., & Hutter, M. (2007). Universal intelligence: A definition of machine intelligence. *Minds and Machines*, *17*(4), 391–444.
Lewis, D. (1981). Causal decision theory. *Australasian Journal of Philosophy*, *59*(1), 5–30.
Mayer, D., Feldmaier, J., & Shen, H. (2016). Reinforcement learning in conflicting environments for autonomous vehicles. In *International workshop on robotics in the 21st century: Challenges and promises*. arXiv: 1610.07089.
Meacham, C. J. G. (2010). Binding and its consequences. *Philosophical Studies*, *149*(1), 49–71. https://doi.org/10.1007/s11098-010-9539-7.
Muehlhauser, L., & Helm, L. (2012). *Intelligence explosion and machine ethics*. Machine Intelligence Research Institute. https://intelligence.org/files/IE-ME.pdf. Accessed 22 Feb 2019.
Nozick, R. (1969). Newcomb's problem and two principles of choice. In N. Rescher, et al. (Eds.), *Essays in honor of Carl G. Hempel* (pp. 114–146). Berlin: Springer.
Oesterheld, C. (2018a). *Doing what has worked well in the past leads to evidential decision theory*. https://casparoesterheld.files.wordpress.com/2018/01/learning-dt.pdf. Accessed 22 Feb 2019.
Oesterheld, C. (2018b). Newcomb's problem, the Prisoner's dilemma and large universes: A consideration for consequentialists. In *Talk at the 15th conference of the international society for utilitarian studies*. Karlsruhe Institute of Technology (KIT), July 24–26, 2018.
Pearl, J. (2009). *Causality. Models, reasoning, and inference*. Cambridge: Cambridge University Press.
Piccione, M., & Rubinstein, A. (1997). On the interpretation of decision problems with imperfect recall. *Games and Economic Behavior*, *20*, 3–24.
Poellinger, R. (2013). *Unboxing the concepts in Newcomb's paradox: Causation, prediction, decision*. http://philsci-archive.pitt.edu/9887/7/newcomb_in_ckps.pdf. Accessed 22 Feb 2019.
Price, H. (1986). Against causal decision theory. *Synthese*, *67*, 195–212.
Price, H. (2012). Causation, chance, and the rational significance of supernatural evidence. *Philosophical Review*, *121*(4), 483–538.
Price, H., & Corry, R. (Eds.). (2007). *Causation, physics, and the constitution of reality: Russell's republic revisited*. Oxford: Oxford University Press.
Ross, S. M. (2007). *Introduction to probability models* (9th ed.). Cambridge: Academic Press.
Russell, S., & Norvig, P. (2010). *Artificial intelligence. A modern approach* (3rd ed.). London: Pearson Education, Inc.
Skyrms, B. (1982). Causal decision theory. *The Journal of Philosophy*, *79*(11), 695–711.
Soares, N. (2014a). *Newcomblike problems are the norm*. http://mindingourway.com/newcomblike-problems-are-the-norm/. Accessed 22 Feb 2019.

Soares, N. (2014b). *Why Ain't you rich?*https://intelligence.org/2014/10/07/nate-soares-talk-aint-rich/. Accessed 22 Feb 2019.

Soares, N, & Fallenstein, B. (2014a). *Aligning superintelligence with human interests: A technical research agenda*. Technical report. 2014-8. Machine Intelligence Research Institute. https://intelligence.org/files/TechnicalAgenda.pdf. Accessed 22 Feb 2019.

Soares, N, & Fallenstein, B. (2014b). *Toward idealized decision theory*. Technical report 2014-7. Machine Intelligence Research Institute. arXiv: 1507.01986.

Soares, N., & Levinstein, B. A. (2017). Cheating death in damascus. In *Formal epistemology workshop (FEW) 2017*. University of Washington, Seattle, USA. https://intelligence.org/files/DeathInDamascus.pdf. Accessed 22 Feb 2019.

Sorg, J. D. (2011). *The optimal reward problem: Designing effective reward for bounded agents*. PhD thesis, University of Michigan. https://deepblue.lib.umich.edu/bitstream/handle/2027.42/89705/jdsorg_1.pdf. Accessed 22 Feb 2019.

Spohn, W. (2003). Dependency equilibria and the causal structure of decision and game situation. *Homo Oeconomicus*, *20*, 195–255.

Spohn, W. (2012). Reversing 30 years of discussion: Why causal decision theorists should one-box. *Synthese*, *187*(1), 95–122.

Sutton, R. S., & Barto, A. G. (1998). *Reinforcement learning: An introduction*. Cambridge: MIT Press.

Treutlein, J. (2018). How the decision theory of Newcomb like problems differs between humans and machines. In *Talk at the 2nd workshop on decision theory & the future of artificial intelligence in Munich, Germany*.

Treutlein, J., & Oesterheld, C. (2017). *A wager for evidential decision theory*. Unpublished manuscript.

Wedgwood, R. (2013). Gandalf's solution to the Newcomb problem. *Synthese*, *190*(14), 2643–2675. https://doi.org/10.1007/s11229-011-9900-1.

Weirich, P. (2016). Causal decision theory. In *The Stanford encyclopedia of philosophy*. Spring 2016.

Yudkowsky, E. (2010). *Timeless decision theory*. The Singularity Institute. http://intelligence.org/files/TDT.pdf. Accessed 22 Feb 2019.

Synthese (2021) 198 (Suppl 27):S6505–S6527
https://doi.org/10.1007/s11229-019-02235-4

S.I.: DECTHEORY&FUTOFAI

Causal concepts and temporal ordering

Reuben Stern[1]

Received: 29 March 2018 / Accepted: 29 April 2019 / Published online: 3 May 2019

Abstract

Though common sense says that causes must temporally precede their effects, the hugely influential interventionist account of causation makes no reference to temporal precedence. Does common sense lead us astray? In this paper, I evaluate the power of the commonsense assumption from within the interventionist approach to causal modeling. I first argue that if causes temporally precede their effects, then one need not consider the outcomes of interventions in order to infer causal relevance, and that one can instead use temporal and probabilistic information to infer exactly when X is causally relevant to Y in each of the senses captured by Woodward's interventionist treatment. Then, I consider the upshot of these findings for causal decision theory, and argue that the commonsense assumption is especially powerful when an agent seeks to determine whether so-called "dominance reasoning" is applicable.

Keywords Causation · Causal Bayes nets · Probabilistic causation · Interventions · Causal decision theory · Dominance reasoning

1 Introduction

When I told my father that I was writing a paper about why it is useful to assume that causes temporally precede their effects, he scoffed. "Why on Earth would anyone waste their time defending such a position?" … "That's just part of what it means for something to cause something else!"[1] My father is not alone in treating causation this way. There are some philosophers (e.g., Hume) who share his opinion and there are also many empirical studies (e.g., Lagnado and Sloman 2006; Burns and McCormack 2009; Bramley et al. 2017) that suggest that human beings, as a species, treat causes as though they necessarily temporally precede their effects.

Chapter 6 was originally published as Stern, R. Synthese (2021) 198 (Suppl 27): S6505–S6527. https://doi.org/10.1007/s11229-019-02235-4.

[1] For what it's worth, my father is no stranger to thinking about causation. He is a lawyer, and is sometimes tasked with establishing causation.

✉ Reuben Stern
reuben.stern@gmail.com

1 Kansas State University, 1116 Mid Campus Dr North, 201 Dickens Hall, Manhattan, KS 66506-0803, USA

But many contemporary philosophical treatments of causation say that most human beings (including my father and Hume) are wrong. Though it is popular for contemporary philosophers to grant that causes (in some sense) almost always temporally precede their effects, it is less popular to maintain that causes *must* (as a matter of conceptual necessity) temporally precede their effects.[2] The reasons for this are complex. Some are convinced by Dummett (1954, 1964) that we can imagine scenarios in which retro-causation is present, while others are convinced by Price (1996) that no imagination is required—i.e., that there are *actual* (albeit somewhat peculiar) instances of retro-causation in the real world.[3] And there are still others who make no claims about whether there are (possible or actual) cases of retro-causation, but who think that causal relevance is best analyzed in terms that make no reference to temporal precedence.

Woodward (2003) is an example of the third kind. According to his hugely influential interventionist treatment of causation, we can understand whether, and in what sense, X is causally relevant to Y without knowing anything about whether X temporally precedes Y. Though the interventionist account of causation is popular for many reasons, one of its most impressive features is its ability to distinguish between several kinds of causal (ir)relevance that typically elude Humean accounts. For example, unlike one of Hume's greatest defenders, Spohn (2012), Woodward neatly captures the sense in which X can be causally relevant to Y even when X is not a *total* cause of Y, and aptly diagnoses failures of transitivity—i.e., cases where X is causally relevant to Y, Y is causally relevant to Z, while X is not causally relevant to Z.

In this paper, my aim is to illustrate that the Humean, too, can distinguish between these kinds of causal (ir)relevance if she assumes that causes must temporally precede their effects, and, therefore, that the commonsense assumption is more powerful than the current philosophical landscape would suggest. More specifically, I argue that if causes temporally precede their effects, then one need not consider the outcomes of interventions in order to infer causal relevance, and one can instead use temporal and probabilistic information to infer exactly when X is causally relevant to Y in each of the senses captured by Woodward's (2003, 2008) interventionist treatment.

If my argument is successful, it does not mean that Dummett and Price are wrong to think that retro-causation is possible, or that my father is right to insist on the conceptual connection between cause and time. Rather, it means that in the great many environments where causes do, as a matter of fact, temporally precede their effects, the commonsense assumption makes it possible to infer whether, and in what sense, X is causally relevant to Y without having to consider what would have happened were someone to have intervened. Since considering the outcomes of interventions seems like a complicated mental task, it should be clear that there are many contexts in which it would be useful to infer facts about causal (ir)relevance from just temporal and probabilistic information. But I will also present one specific decision-theoretic context in which it is demonstrably useful—namely, when an agent seeks to determine whether so-called "dominance reasoning" is applicable.

[2] Philosophers who attempt to reduce causal relevance to a kind of probabilistic relevance (e.g., Eells 1991; Spohn 2012) are notable exceptions.

[3] For example, Price (2001) argues that retro-causation is at work in the quantum mechanical context of the Bell correlations.

My aim therefore can be succinctly stated as follows. I want to show that even if my father and the rest of mankind are thick-headed insofar as they think that causes must, as a matter of conceptual necessity, temporally precede their effects, there is good reason for their thick-headedness—namely, it enables them to efficiently determine what is causally relevant to what else, and in what sense, in most of the contexts that they ever actually confront.

2 Woodward's interventionist treatment of causal relevance

As I mentioned above, one of the most impressive features of the interventionist treatment of causation is its ability to distinguish several kinds of causal relevance. What are these kinds of causal relevance? Woodward (2003, 2008) proposes the following three analyses of causal relevance, where X and Y denote variables (i.e., sets of collectively exhaustive and mutually exclusive state-descriptions), and $\mathbf{V}$ denotes a set of variables.

1. X is a **direct cause** of Y relative to $\mathbf{V}$ if and only if there is a possible intervention on X that will change the probability distribution over Y when all other variables in $\mathbf{V}$ are held fixed at some value by interventions.
2. X is a **total cause** of Y if and only if there is a possible intervention on X that will change the probability distribution over Y.
3. X is a **contributing cause** of Y if and only if there is some $\mathbf{V}$ that includes X and Y, relative to which (i) there is a directed path from X to Y and (ii) there is some intervention on X that will change the probability distribution over Y when all of the variables in $\mathbf{V}$ not on this directed path are held fixed at some value by interventions.[4]

In order to understand these notions of causal relevance, and why each one deserves a place in a theory of causation, we must first understand what it is to *intervene* on a variable. Woodward (2003) develops his own detailed understanding of interventions, but it will suffice for our purposes to understand the intervention on X as an exogenous cause of X (i.e., a cause of X that has no causes in $\mathbf{V}$ and that is not a direct cause of any variable in $\mathbf{V}$ other than X) that can be exploited (or conditioned upon) to set the value of X to any of its possible values. Given the Causal Markov Condition,[5] the intervention on X is possibly correlated with variables that are causally downstream from X,

[4] This formulation is slightly different from any of Woodward's, but it's a good approximation of his view. The reason for any discrepancy is that Woodward (2008) modifies his original (2003) view in response to Strevens's (2007) criticism that the original definition of 'contributing cause' was relativized to a variable set, and this is my best guess at what Woodward would have said had he anticipated Strevens's criticism. Woodward's actual reply to Strevens was that "X is a contributing cause *simpliciter* of Y (in a sense that isn't relativized to any particular variable set $\mathbf{V}$) as long as it is true that there exists a variable set $\mathbf{V}$ such that X is correctly represented as a contributing cause of Y with respect to $\mathbf{V}$." Here, because I believe that Woodward's original $\mathbf{V}$-relative notion of 'contributing cause' plays no important role once Woodward's (2008) notion is on the table, I have simply modified Woodward's original definition of 'contributing cause' so that X qualifies as a contributing cause of Y when Woodward (2008) says that X qualifies as a "contributing cause *simipliciter*" of Y.

[5] The Causal Markov Condition is an axiom of the graphical approach to causal modeling and is argued by Hausman and Woodward (1999) to be implicit in any treatment of causation according to which causes can

but cannot be correlated with any variables that are not causally downstream from X.[6] For this reason, it is sometimes helpful to conceptualize the effect of conditioning on an intervention variable in terms of setting the intervened upon variable to a particular value in a way that (i) severs any causal dependencies between the intervened upon variable and its endogenous (non-intervention) parents, but (ii) leaves intact the dependencies between the intervened upon variable and any of its effects.[7]

In order to better understand this notion, it is helpful to consider how it applies to a concrete example that doubles as an introduction to direct causes and total causes.

Suppose that two dominoes (*D1* and *D3*) are positioned such that they can be knocked over by a gust of wind (W) that occurs when someone opens the door. Further suppose that there is another domino (*D2*) that lies beyond the reach of the wind, but that is situated between *D1* and *D3* such that *D1* is blocked from ever making direct contact with *D3*, but also such that *D1* can knock over *D2*, and that *D2* can knock over *D3*. Fig. 1 is an intuitive representation of this scenario.

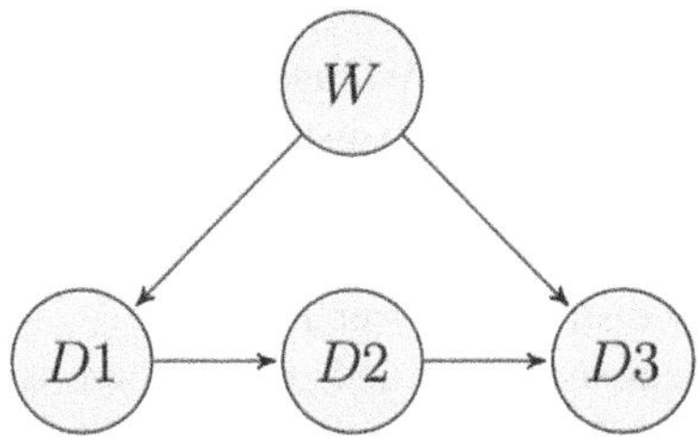

Fig. 1 The dominoes

One way of describing what we find intuitive about this case is that *D1*'s falling seems to indirectly causally promote *D3*'s falling by way of *D2*'s falling, and that whether the wind blows (W) seems to have a direct causal effect on whether *D1* and *D3* fall, and an indirect causal effect on whether *D2* falls (by way of *D1*). Woodward's analyses of 'direct cause' and 'total cause' deliver these results. If we intervene on the value of *D1* by, say, manually tipping over the domino,[8] our action will have an effect on the probabilities of *D2* and *D3*, but not W. This means that *D1* qualifies as a total cause of *D2* and *D3*. Then, because manually tipping over *D1* has no effect on the probability of *D3* when we additionally intervene to "hold fixed" the values of *D2* (e.g., by manually holding the domino in an upright position) and W (e.g., by blocking the air flow), but does affect the probability of *D2* when we intervene to "hold fixed" the values of *D3* and W, Woodward's analysis of 'direct cause' tells us that *D1* is a direct cause of *D2*, but only an indirect cause of *D3*.

Footnote 5 continued
be used to manipulate their effects, including Woodward's (2003) treatment. We will discuss this condition in more detail in the next section of this paper.

6 As we will see later, it does not follow from the Causal Markov Condition that the intervention on X *must* be correlated with any variables that are causally downstream from X.

7 This is how Pearl (2009) understands interventions.

8 This qualifies as an intervention relative to the system at hand since the wind has no effect on whether we manually tip over the domino.

Were every cause of an effect a total cause of its effect, then these two analyses could conceivably tell us everything that we need to know about causal relevance. That is, we could treat X as a cause of Y exactly when X is a total cause of Y, and then use Woodward's analysis of 'direct cause' to distinguish between proximate and remote causes. But alas, it is not true that X is a cause of Y exactly when X is a total cause of Y, hence Woodward's inclusion of the 'contributing cause' category.

Suppose, for example, that taking birth control (BC) promotes thrombosis (TH) if you are pregnant (or if you aren't), but also reduces the risk of thrombosis by reducing the risk of pregnancy (P). Now, further suppose that the probabilistic effect of BC that is mediated by P cancels out the probabilistic effect that is due to BC's direct effect on TH.[9]

In this case (depicted in Fig. 2), when we intervene on BC by administering the pill, there is no probabilistic effect on TH (since the causal dependencies that are left intact cancel each other out), so BC is not a total cause of TH. But of course BC is causally relevant to TH. After all, we know that BC has a deleterious effect on the body (represented by the direct causal relationship between BC and TH). So we cannot just say that X is causally relevant to Y exactly when X is a total cause of Y.

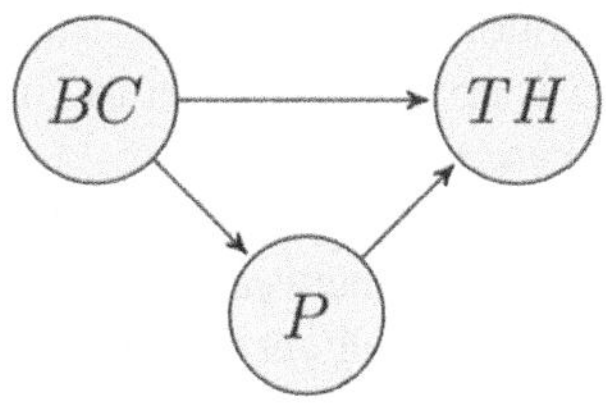

Fig. 2 The birth control case

It is tempting to respond to this case by claiming that X is a cause of Y exactly when there is some $\mathbf{V}$ relative to which there is a directed path between X and Y—or, put differently, exactly when there is a chain of direct causal relationships between X and Y. This gets the above case right (since there are directed paths between BC and TH), but it sometimes posits causal relevance where there is none. This is because this analysis implicitly assumes that causal relevance is transitive, but causal relevance is not actually transitive.

Consider a slight variant of McDermott's (1995) "dog bite" counterexample to the transitivity of causation.[10] In this scenario, a terrorist is contemplating pressing a button that will detonate a bomb and very probably blow up a building, but there is a dog present who threatens to bite the terrorist's right hand. As things turn out, the probability that the terrorist detonates the bomb is not affected by whether the dog bites, but the probability that the terrorist uses her right or left hand (in the event that she pushes the button) is affected. If we allow DB to represent whether the dog bites, P to represent whether the button is pressed and if so with which hand, and

[9] This is a common variant of an example introduced by Hesslow (1976).

[10] The only difference is that there is no local determinism in the model. Zhang and Spirtes (2008) likewise discuss a version of McDermott's case in which there is no local determinism.

E to represent whether the explosion occurs, then Fig. 3 represents the direct causal relationships that Woodward countenances.

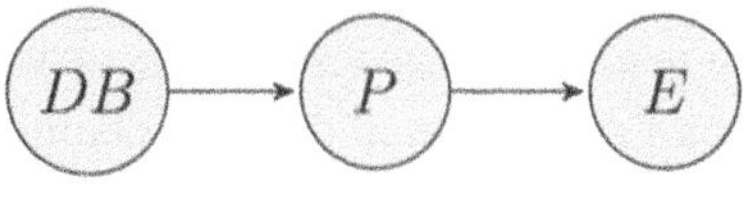

Fig. 3 The dog bite

Here, the intuition is strong that *DB* is not causally relevant to *E* even though there is a directed path from *DB* to *E* in the variable set over which Fig. 3 is defined. That is, even though (i) *DB* exerts causal influence over *P* (insofar as it causally influences which hand the terrorist uses, if she pushes the button), and (ii) *P* exerts causal influence over *E* (insofar as whether there is an explosion depends on whether the terrorist pushes the button), it does not seem that *DB* causally influences *E* (since there is no sense in which whether there is an explosion depends on whether the dog bites).

Since *DB* is not a cause of *E*, and *BC* is not a total cause of *TH*, we need some notion of causal relevance that is neither captured by the transitive closure of 'direct cause',[11] nor captured by Woodward's notion of 'total cause'.[12] In order to see that Woodward's notion of 'contributing cause' is up to the task, let us consider its application to both the Fig. 2 case and the Fig. 3 case.

In order to check whether *BC* qualifies as a contributing cause of *TH*, we must first check whether there is a directed path between *BC* and *TH* in some variable set. As things turn out, if we consider the variable set depicted in Fig. 2, there are two directed paths between *BC* and *TH*. If *P* is held fixed at either value, interventions on *BC* are correlated with *TH*, so there is a direct causal relationship here. Similarly, if *TH* is held fixed at either value, interventions on *BC* are correlated with *P*, so there is a direct causal relationship here. And if *BC* is held fixed at either value, interventions on *P* are correlated with *TH*, so there is yet another direct causal relationship. This means that Woodward attributes every direct causal relationship that is captured in Fig. 3.

But in order for *BC* to count as a contributing cause of *TH*, it's not enough that there is a directed path between *BC* and *TH* in some **V**. There must also be some intervention on *X* that will change the probability distribution over *Y* when all of the variables in **V** not on this directed path are held fixed at some value by interventions. In this case, it is clear that there are such interventions. For example, when we hold fixed the value of *P* by intervention (e.g., by impregnating someone through in vitro fertilization), interventions on *BC* have an effect on *TH*, and *BC* therefore qualifies as a contributing cause of *TH*.

---

[11] Spohn (2012) takes this option, but limits the domain of his theory to dichotomous variables. (His theory therefore does not apply to the "dog bite" because *DB* has three values.) Spohn convincingly argues that causation *is* transitive when the variables are dichotomous, but like Woodward, we are interested in developing an account of causation that extends to settings where the variables are not dichotomous.

[12] Though *DB* is not a total cause of *E*, we cannot say that *X* is a cause of *Y* exactly when *X* is a total cause of *Y* because this gets the Fig. 2 case wrong. Similarly, though there is a causal chain from *BC* to *TH*, we cannot say that *X* is a cause of *Y* exactly when *X* there is a directed path from *X* to *Y* because this gets the Fig. 3 case wrong.

Even though there is likewise a directed path between *DB* and *E* (because Woodward attributes direct causal relevance exactly where Fig. 3 does), Woodward does not get the result that *DB* is a contributing cause of *E*. This is because there is nothing along some other directed path on which we can intervene in order to reveal a dependency between *DB* and *E*. That is, even if we considered variable sets richer than this one, we know that no matter what we hold fixed, there will be no dependence between interventions on *DB* and *E*. This is roughly because the lack of dependence between interventions on *DB* and *E* occurs along a single directed path, rather than along multiple directed paths. Thus we cannot unveil a dependency between *DB* and *E* by holding fixed a variable at some value along one of multiple directed paths between *DB* and *E*.

It is now clear that Woodward's notion of 'contributing cause' is up to the task at hand. That is, one can reasonably contend that *X* is a cause of *Y* exactly when *X* is a contributing cause of *Y* because (i) every total cause counts as a contributing cause, (ii) causes that do not qualify as total causes (because there is path cancellation) still qualify as contributing causes, and (iii) it is not the case that *X* counts as a contributing cause of *Y* whenever there is an intransitive directed path between *X* and *Y*. Woodward's interventionist treatment of causation therefore allows one to distinguish between different kinds of causal relevance without forgoing our intuitions about causal relevance.

3 Causation without intervention

Woodward's interventionist treatment of causation is compelling insofar as it distinguishes between these three kinds of causal relevance, but it is sometimes hard to determine the extension of Woodward's account because it is sometimes difficult to determine the outcomes of interventions. Suppose, for example, that you are curious whether the widowhood effect is real—i.e., whether your spouse's death exerts causal influence on your life expectancy.[13] Woodward's analyses suggest that we can determine this simply by considering whether there is an effect on the probability that you live to a certain age in the event that we intervene on whether you are a widow (perhaps while holding fixed the values of some other variables by intervening). But these interventions are hard to perform. We cannot experimentally implement these interventions because we cannot kill spouses, and we also are not very good at thinking about what their effects would be were we to implement them. Would killing my wife have an effect on my life expectancy? I have absolutely no idea.

In settings where it is difficult to carry out actual interventions, or to mentally simulate hypothetical interventions, it is clear that it would be useful to use some other kind of information to distinguish between kinds of causal relevance. Until now, to my knowledge, there has been no account of causation other than the interventionist account that has been able to effectively draw these distinctions between kinds of causal relevance. For example, Pearl (1988) and Spohn (2001, 2012) show that tem-

[13] Elwert and Christakis (2008) use graphical causal models to show that the widowhood effect is real without killing anyone.

poral information (together with probabilistic information) can be used to underwrite powerful causal inference, but neither goes as far as to show that to use this kind of information to draw distinctions as rich as Woodward's.

Here, I want to investigate the possibility of proposing three analyses of causal relevance that agree with Woodward's interventionist analyses in settings where causes temporally precede their effects, but that make no reference to interventions.[14] Since these analyses are premised upon the claim that causes must temporally precede their effects, these analyses are guaranteed to agree with Woodward's interventionism only in settings where the variables in **V** are temporally ordered,[15] and only when there is no retro-causation or simultaneous causation. But since human beings encounter a great many contexts in which they're aware of the temporal order in which the causal factors under consideration are settled, and in which there is no retro-causation or simultaneous causation, these analyses should demonstrate the considerable benefits that can be accrued (from the interventionist's perspective) by assuming that causes must temporally precede their effects.

Before proceeding to develop these analyses, some more table setting is required. Specifically, we must limit the domain not only to contexts where the variables under consideration are temporally ordered, but also to contexts (i) where the relevant variable sets are causally sufficient (in the sense that **V** does not omit any confounding common causes of variables included within **V**)[16] and (ii) where the relevant probability distributions over these variables are positive (in the sense that every assignment of values over **V** is assigned non-zero probability). Though it may seem that these restrictions are somewhat strict (especially since Woodward doesn't frequently draw our attention to them), they are reasonable in the present context for several reasons. First, if Hausman and Woodward (1999) are right that interventionism presupposes the Causal Markov Condition, then the variable sets at play need to be causally sufficient in order to deliver reasonable results.[17,18] Second, it is often thought that the

[14] Some may quibble with my choice to refer to these sets of necessary and sufficient conditions as *analyses* because their right-hand sides include causal notions that are too similar to the notions on the left (e.g., causal sufficiency). If you're one of these people, please feel free to replace every mention of "analysis" with a term that you deem more appropriate. (I do not mean to use the term in a loaded way.).

[15] This is not a trivial restriction. For example, when dealing with type-level variables like income or intelligence, the variables often do not specify what happens at a specific time, and it is therefore difficult to recover a temporal order. When this is impossible, these analyses are silent.

[16] It is worth noting that this does *not* mean that **V** must include *every* common cause of any variables included within **V**. For example, the agent can leave out distal common causes of X and Υ in the event that she has included more proximate common cause of X and Υ that screens off the distal cause. For ease of exposition, I do not specify exactly which common causes are confounding in the sense that matters for causal sufficiency, but there is at least one juncture of the paper (flagged in footnote 40) at which this is relevant.

[17] The Causal Markov Condition entails that when two variables, X and Y, are correlated, then either (i) X and Y share a common ancestor, (ii) there is a directed path from X to Y, or (iii) there is a directed path from Y to X. Thus when the Causal Markov Condition is assumed of a variable set that excludes a common cause of two variables, X and Y, we must posit a directed path between X and Y even when there isn't one.

[18] Though Hausman and Woodward argue that the CMC is implicit in interventionism, it is important to note that we can often discover causal relationships by intervening on a variable without considering any causally sufficient variable sets. Consider randomized controlled trials. If the experimental manipulation on X is genuinely random (as it is in the ideal case), then we can infer whether X is a total cause of Y

relata of causal relationships must be distinct (or "modular"),[19] and there is a natural sense in which the variables that comprise **V** qualify as distinct when the probability distribution over **V** is positive—namely, every variable can take any of its values, no matter what the values of the other variables are. Finally, and perhaps most importantly, the great majority of probability distributions that humans encounter in the context of causal inference are positive. This is because human beings almost never attend to enough causal detail to render any values of any variables incompatible with one another.[20]

We must also say more about the axioms of the graphical approach to causal modeling that are (explicitly or implicitly) assumed by interventionists in this domain.[21] This is because we will not be able to understand how the temporal counterparts of Woodward's analyses deliver the same results (in contexts where causes temporally precede their effects) without understanding the implications of these axioms—namely, the Causal Markov Condition and the Causal Minimality Condition (or Minimality, for short).

Pearl (2009, pp. 16–17) neatly summarizes implications of the Causal Markov Condition in the graphical terms of ***colliders*** and ***d-separation***.

collider: A variable is a *collider* along a path if and only if it is the direct effect of two variables along the path. (So M is a collider along $I \rightarrow M \leftarrow J$ but not along $I \leftarrow M \rightarrow J$ or $I \rightarrow M \rightarrow J$.)

d-separation: A path between two variables, X and Y, is *d-separated* (or blocked) by a (possibly empty) set of variables, Z, if and only if

i. the path between X and Y contains a non-collider that is in Z, or
ii. the path contains a *collider*, and neither the collider nor any descendant of the collider is in Z.

Given the CMC, if every (undirected) path between a pair of variables in **V** is *d-separated* by Z (according to a given DAG), then the pair of variables must be probabilistically independent of each other conditional on any assignment of values over Z. This means, for example, that if Fig. 1 is the correct representation of the domino system, then *D1* and *D3* must be probabilistically independent given any assignment of values over *D2* and *W* (because *D2* and *W* are non-colliders on the only paths between *D1* and *D3*, and thus block every path between *D1* and *D3*). On the other hand, the CMC does *not* entail that *D2* and *W* are probabilistically independent

Footnote 18 continued
by determining whether the probability of Y changes when we experiment on X. This is one important advantage that Woodward's interventionist analyses have over the temporal analyses that I propose here. See Eberhardt (2014) for extensive discussion of these issues.

19 Hausman and Woodward (1999) speak in terms of modularity.

20 Consider the dominoes from Fig. 1. Under normal circumstances, this is about as close as we get to considering a causal system in which there is local determinism, but even here, it is clear that every assignment of values over **V** should be assigned non-zero probability since, e.g., each of the dominoes *can* fall when the others do not. After all, we already mentioned that humans *can* manipulate whether each of the dominoes is upright without interfering with the status of the other dominoes.

21 As I mentioned before, Hausman and Woodward (1999) convincingly argue that interventionism presupposes the Causal Markov Condition. Zhang and Spirtes (2011) make the same point about the Causal Minimality Condition when the probability distribution over **V** is positive.

given any assignment of values over *D1* and *D3* because *D3* is a collider along the path $D2 \rightarrow D3 \leftarrow W$.

The CMC allows us to infer when the variables represented in a DAG must be probabilistically independent, and when the variables over which a probability distribution is defined must be causally connected, but it never allows us to infer when variables are probabilistically *dependent*, or when variables are *not* causally connected.[22] In order to make these additional inferences, some other condition must be posited in addition to the CMC. The most discussed of these conditions is perhaps the Causal Faithfulness Condition (CFC), which is satisfied by a graph and probability distribution if and only if there are no conditional independencies in the probability distribution that are not implied by the graph, given the CMC. Put differently, the CFC entails that if some path between a pair of variables in **V** is *d-connected* (equivalently, not d-separated) by *Z*, then the pair of variables must be probabilistically *dependent* of each other conditional on some assignment of values over *Z*.

The CFC is unreasonable to assume in the current setting because the causal structures depicted in Figs. 2 and 3 are not faithful to the probability distributions with which they are paired. In order to see this, we only need to remember that *BC* and *TH* (in Fig. 2) and *DB* and *E* (in Fig. 3) are unconditionally probabilistically independent but d-connected by the empty set of variables.[23] Does this spell doom for probabilistic treatments of causality? The answer is no. We can replace the oft-discussed CFC with a condition that is strictly weaker than the CFC, and that is satisfied by the examples under consideration. This condition, known as Minimality (MIN), is satisfied by a DAG and probability distribution if and only if there exists no proper subgraph of the DAG that satisfies the CMC with respect to the probability distribution.[24] Unlike the CFC, MIN is sometimes treated as an axiom of the graphical approach to causal modeling,[25] and has been shown by Zhang and Spirtes (2011) to be implicit in Woodward's interventionist treatment of causation when the probability distribution over **V** is positive. Moreover, it is clear that MIN is satisfied by the examples depicted in Figs. 2 and 3. In Fig. 2, if we delete the arrow between *BC* and *TH*, then the CMC is violated because *BC* and *TH* are correlated given any value of *P*, and if we delete the arrow between *BC* and *P* or the arrow between *P* and *TH*, the CMC is again violated because both pairs of variables (i.e., *BC* and *P* and *P* and *TH*) are unconditionally probabilistically dependent. In Fig. 3, if we delete the arrow between *DB* and *P* or the arrow between *P* and *E*, then the CMC is violated because both pairs of variables (i.e., *DB* and *P* and *P* and *E*) are unconditionally probabilistically dependent.

---

[22] For example, even though the CMC does not entail that *D2* and *W* are probabilistically independent given any assignment of values over *D1* and *D3*, it does *not* entail that *D2* and *W* are probabilistically *dependent* given any assignment of values over *D1* and *D3*.

[23] Throughout this paper, I sometimes speak in terms of variables (rather than paths) being d-separated. When I do so, I follow Pearl (2009) by referring to *X* and *Y* as d-separated by *Z* when every path between *X* and *Y* is d-separated by *Z*.

[24] A DAG, *G*, qualifies as a proper sub-DAG of G^* if and only if (i) *G* contains fewer arrows than G^*, and (ii) every arrow in *G* is oriented in the same direction as its corresponding arrow in G^*.

[25] See Spirtes et al. (2000).

As Hitchcock (2018) explains, Pearl (1988) proves that:
If

i. the variables in **V** are time-indexed, such that only earlier variables can cause later ones;
ii. the probability P assigns positive probability to every possible assignment of values of the variables in **V**;
iii. [**V** is causally sufficient];[26]
iv. and the probability measure P satisfies the Markov and Minimality Conditions with respect to the DAG;

then it is possible to uniquely identify the DAG on the basis of P.

If the inclusion of an arrow in a DAG is taken to represent the existence of a direct causal relationship, Pearl's result shows us that it is possible to identify what is directly causally relevant to what else in a sense that agrees with the interventionist analysis of 'direct cause' without checking the outcomes of any interventions. More specifically, when causes temporally precede their effects, Pearl shows that if (i) **V** is causally sufficient, (ii) the variables comprising **V** are temporally ordered,[27] and (iii) the probability distribution defined over **V** is positive, then, given the CMC and MIN, there is just one DAG that is compatible with the given probability distribution. Thus one can infer the true DAG—i.e., the DAG championed by interventionism (given its assumption of the CMC and MIN)—from just a probability distribution that is defined over a temporally ordered set of variables.

Since DAGs represent what is directly causally relevant to what else, this implies that there must be some way to analyze 'direct cause' that makes no reference to interventions or their outcomes, but that yields the same extension as Woodward's analysis in the settings under consideration. But it is not yet clear what this conception of 'direct cause' is, nor is it clear whether probabilistic and temporal information can be used to determine when X is a contributing cause of Y, or when X is a total cause of Y (since DAGs are just representations of what is directly causally relevant to what else). Let us deal with these issues in turn. I submit that the following temporal analysis of 'direct cause' is provably co-extensive with Woodward's in the settings under consideration when causes temporally precede their effects.

1*. X is a **direct cause** of Y relative to **V** if and only if (i) X is temporally prior to Y and (ii) X and Y are probabilistically dependent conditional on some assignment of values over all and only the variables in **V** that is/are temporally prior to Y excepting X.[28]

This analysis delivers the correct (interventionist) results because the set of variables that are temporally prior to Y excepting X provably d-separates any indirect path between X and Y.[29] Thus if a correlation persists between X and Y under some

[26] Hitchcock puts this condition differently, but the content is the same for all intents and purposes.

[27] The temporal ordering must be a complete ordering, rather than a partial ordering. Temporal orderings are complete orderings because they are opinionated (in the sense that the value of X is settled either temporally before, temporally after, or at the same time as the value of Y).

[28] Spohn's (1980) treatment of 'direct cause' is very similar to this one.

[29] Proof: Suppose that causes temporally precede their effects and that X temporally precedes Y. (1) Every indirect path between X and Y that is d-connected by the empty set must include an ancestor of Y (either a

assignment of values to this set of variables, then there must be an arrow between X and Y in order for the CMC to be satisfied. And if no such correlation persists, then MIN entails that there is no such arrow between X and Y (because the DAG would satisfy the CMC were this arrow excluded).[30]

One can also verify that this analysis agrees with Woodward's by checking how it fares when applied to the examples from the last section. For example, if we consider the domino case, 1* vindicates the graphical representation represented in Fig. 1. Under the assumption that causes temporally precede their effects, it is clear that W is settled prior to $D1$, that $D1$ is settled prior to $D2$, and that $D2$ is settled prior to $D3$. Thus, provided that the probability distribution over $\mathbf{V}$ is positive, it is clear (i) that W is correlated with $D1$ (and, therefore, that 1* entails that $W \rightarrow D1$), (ii) that $D1$ is correlated with $D2$ given any value of W (and, therefore, that 1* entails that $D1 \rightarrow D2$), (iii) that $D2$ is correlated with $D3$ given any assignment of values over W and $D1$ (and that 1* therefore entails that $D2 \rightarrow D3$), and (iv) that W is correlated with $D3$ given any assignment of values over $D1$ and $D2$ (and, therefore, that 1* entails that $W \rightarrow D3$). Moreover, 1* correctly entails that W is not a direct cause of $D2$ because W is not correlated with $D2$ given any value of $D1$.[31,32]

The lesson here is that when causes temporally precede their effects, 1* says that X directly causes Y relative to $\mathbf{V}$ exactly when interventions on X yield a change in the probability distribution over Y while holding fixed every other variable in $\mathbf{V}$ at some value. Thus we have established that we can sometimes use temporal and probabilistic information to infer when (and when not) interventions on X bring about a change in the probability distribution over Y.

Is there some way to use this finding to develop temporal analyses of contributing causal relevance and total causal relevance? Since DAGs do not straightforwardly represent whether variables are relevant in either of these senses, this is somewhat trickier business, and it is correspondingly harder to prove definitive results. But I submit that there are temporal treatments of both kinds of causal relevance (developed in what follows) that single out the same kinds of relevance that are singled out by

Footnote 29 continued
common ancestor of X and Y, or an ancestor of Y that is a descendant of X), and the set of variables temporally prior to Y excepting X must therefore d-separate these paths (since this set includes these ancestors). (2) Every indirect path between X and Y that is d-separated by the empty set must include a collider between X and Y that is either a descendant of Y or a descendant of some ancestor of Y (because these are the only ways for there to be a collider between X and Y) and the set of variables temporally prior to Y excepting X cannot d-connect any of the paths on which these colliders reside since it cannot include any colliders that are descendants of Y (or their descendants) and must include a common ancestor of any other collider and Y. (3) If the set of variables temporally prior to Y excepting X does not d-connect any indirect paths between X and Y that are d-separated by the empty set, and d-separates every indirect path between X and Y that is d-connected by the empty set, then this set of variables d-separates every indirect path between X and Y.

30 When the probability distribution over $\mathbf{V}$ is not positive, there are sometimes multiple DAGs that are compatible with the temporal ordering and that satisfy the CMC and MIN. In these aberrant circumstances, 1* says that X is a direct cause of Y when there is an arrow from X to Y in *every* such DAG.

31 W *is* correlated with $D2$ given some assignments of values over $D1$ and $D3$, but 1* disallows $D3$ from being included in the conditioning set that is used to verify whether $W \rightarrow D2$ because $D3$ is settled after $D2$.

32 It is easy to show that 1* likewise vindicates the graphical representations found in Figs. 2 and 3, but this is left as an exercise for the reader.

Woodward's analyses, and that deliver the same results as Woodward's when applied to the examples that motivate Woodward's own analyses.

Let us confront total causal relevance first. Recall that *X* is a total cause of *Y* exactly when there are interventions on *X* that are correlated with *Y*. When we intervene on *X*, we control for any spurious correlations between *X* and *Y* that are due to common causes, but we do nothing to distinguish between proximate and remote causal relevance, and do not isolate specific causal paths from *X* to *Y*. This means that in order to develop a temporal analysis of 'total cause' that agrees with Woodward's, we need some way to control for spurious correlations due to common causes that, unlike 1*, does *not* block every indirect path from *X* to *Y*. The following analyses of 'total cause' both accomplish this feat.

2*. *X* is a **total cause** of *Y* if and only if *X* is a direct cause of *Y* relative to some causally sufficient variable set containing *X* and *Y* that excludes any variables temporally in between *X* and *Y*.[33]

2**. *X* is a **total cause** of *Y* if and only if (i) *X* is temporally prior to *Y*, and (ii) *X* and *Y* are probabilistically dependent conditional on some assignment of values over *everything* temporally prior to *X*.

Like Woodward's analysis of 'total cause', neither 2* nor 2** is relativized to a variable set. That is, even though 2* says that whether *X* is a total cause of *Y* depends on whether *X* is a direct cause of *Y* relative to a particular variable set, it says that *X* either is or is not a total cause of *Y*, full stop. So both analyses potentially capture the sense in which whether *X* is a total cause of *Y* just depends on the causal relationships that are out there in the world, and not on the particular variable set to which we attend.

It is also easy to see that neither 2* nor 2** distinguishes between proximate and remote causal relevance, or isolates specific paths between *X* and *Y*. In the case of 2*, this is because any potential causal intermediaries are excluded from the variable sets that settle whether *X* is a total cause of *Y*, since, provided that causes temporally precede their effects, any such variables must be temporally between *X* and *Y*. Similarly, in the case of 2**, whether *X* is a total cause of *Y* is settled by checking whether *X* is correlated with *Y* conditional on some assignment of values over everything temporally prior to *X*, and this conditioning set must exclude any potential causal intermediaries (because any such intermediaries cannot be temporally prior to *X* if causes temporally precede their effects).

Both analyses also effectively control for spurious correlations due to common causes. In order to apply 2*, we check whether *X* is a direct cause of *Y* relative to a causally sufficient variable set that excludes any potential causal intermediaries, where the meaning of 'direct cause' is supplied by 1*. Thus 2* inherits 1*'s ability to control for these spurious correlations. In order to apply 2**, we do not check whether *X* is a direct cause of *Y* relative to some particular causally sufficient variable set, but rather just ask whether *X* is correlated with *Y* conditional on some assignment of values over *every* variable that is temporally prior to *X*. This set of variables provably d-separates

[33] Here, 'direct cause' should be understood in terms of 1*.

any paths between *X* and *Y* that are not directed paths from *X* to *Y* (and therefore blocks any spurious dependencies due to common causes).[34]

2* and 2** should thus agree with Woodward's analysis of 'total cause' when applied to the particular cases from the last section.[35] If we apply 2* to the case represented by Fig. 1, we find that *D1* is a total cause of *D3* even though *D1* is not a direct cause of *D3* relative to the variable set over which Fig. 1 is defined. Why? If we consider any variable set that includes *D1* and *D3*, and their common cause, *W*, but that excludes any variables temporally in between *D1* and *D3* (e.g., *D2*), then, relative to that variable set, *D1* will be a direct cause of *D3*. Similarly, if we apply 2* to the example represented by Fig. 2, we (correctly) find that *BC* is not a total cause of *TH* because *BC* is not a direct cause of *TH* relative to any variable set that excludes *P* (as well as any other causal intermediaries). Finally, if we apply 2* to the example represented by Fig. 3, we find that *DB* is not a total cause of *E* because *DB* is not a direct cause of *E* when *P* is excluded. The same goes for 2**'s application to these cases. 2** says (i) that *D1* is a total cause of *D3* because *D1* and *D3* are correlated given some assignment of values over everything preceding *D1* (including *W*),[36] (ii) that *BC* is not a total cause of *TH* because *BC* and *TH* are not correlated given some assignment of values over everything preceding *BC* (because the paths still cancel each other out), and (iii) that *DB* is not a total cause of *E* (because any directed path between *DB* and *E* does not transmit probabilistic dependence.)

The lesson of these analyses is that there are interventions on *X* that change the probability distribution over *Y* (without controlling for causal intermediaries) exactly when, given that causes temporally precede their effects and that *X* temporally precedes *Y*, there are variable sets excluding any potential intermediaries relative to which *X* is

[34] Proof: Suppose that causes temporally precede their effects and that *X* temporally precedes *Y*. (1) Every path between *X* and *Y* that (i) is not a directed path from *X* to *Y* and (ii) is d-connected by the empty set must include an ancestor of *X*, and the set of variables temporally prior to *X* must therefore d-separate these paths (since this set includes these ancestors). (2) Every path between *X* and *Y* that is d-separated by the empty set must include a collider between *X* and *Y* that is either a descendant of *X* or a descendant of some ancestor of *X* (because these are the only ways for there to be a collider between *X* and *Y*) and the set of variables temporally prior to *X* cannot d-connect any of the paths on which these colliders reside since it cannot include any colliders that are descendants of *X* (or their descendants) and must include a common ancestor of any other collider and *X*. (3) If the set of variables temporally prior to *X* does not d-connect any paths between *X* and *Y* that are d-separated by the empty set, and d-separates every path between *X* and *Y* that (i) is not a directed path from *X* to *Y*, and (ii) is d-connected by the empty set, then this set of variables d-separates any paths between *X* and *Y* that are not directed paths from *X* to *Y*.

[35] An anonymous referee points helpfully out that if the laws of nature are deterministic, then 2* and 2** may disagree with each other, thereby making it the case that 2* and 2** can't both agree with Woodward's analysis. The basic idea is that if the laws of nature are deterministic, then if we condition on any possible past, the value of *X* will be determined, and therefore will not be a total cause of anything (since *X* cannot co-vary with anything if *X* does not vary at all). This is worth considering more—especially since (i) it provides a novel way to vindicate Russellian eliminativism about causation and (ii) it is not yet clear how 2* fares in deterministic contexts like these. But the best we can do for now is to bracket these issues by limiting the domain to contexts in which the probability distribution over the variables at hand is positive.

[36] It is tempting to think that conditioning on everything prior to *D1 determines* whether the domino falls (especially if the world is deterministic), and that *D1* therefore can't be correlated with *anything* in this setting. It's thus important to remember that we've restricted the domain of applicability to settings where the probability distribution over the variables at play is positive, and therefore ruled out this kind of determinism.

a direct cause of *Y*, or, alternatively, exactly when *X* and *Y* are correlated given some assignment of values over *every* variable temporally prior to *X*.

2* and 2** are both useful insofar as they provide means for checking whether *X* is a total cause of *Y* without checking the outcomes of any interventions, but the exact senses in which 2* and 2** are useful can come apart. The methodological advice supplied by 2** is easy-to-follow (since it just says that you should check whether *X* and *Y* are correlated whenever you condition on everything that you can think of prior to *X*), but applying 2* involves checking *less* than applying 2** (since you needn't consult a probability distribution defined over *everything* temporally prior to *X*, and can instead limit your focus to just the common causes of *X* and *Y*).[37] Either way, when causes temporally precede their effects, it is clear that we can use probabilistic and temporal information to identify whether *X* is a total cause of *Y* without checking the outcomes of any interventions.

In order to complete our treatment of causal relevance, an analysis of 'contributing cause' is required. This category of causal relevance is of special importance insofar as *X* is supposed to be a contributing cause of *Y* exactly when *X* is a cause of *Y*, period. This means that, in order for the operative notion of 'contributing cause' to effectively perform its job, it must diagnose failures of transitivity (like the one between *DB* and *E* in Fig. 3) while capturing the sense in which *X* can qualify as a cause of *Y* both when *X* is a remote cause of *Y* (e.g., that *D1* is a cause of *D3*) and when *X* is not a total cause of *Y* (e.g., that *BC* is a cause of *TH*).

3*. *X* is a **contributing cause** of *Y* if and only if *X* is a direct cause of *Y* relative to some causally sufficient variable set containing *X* and *Y*.[38]

3* says that *X* is a contributing cause of *Y* exactly when there is *some* variable set relative to which *X* is a direct cause of *Y*. This successfully diagnoses failures of transitivity since there are some pairs of variables that, relative to some variable sets that include them, are connected by a directed path, but that are never connected by a single arrow (no matter what variable set we consider). For example, even though $DB \rightarrow P \rightarrow E$, *DB* does not qualify as a contributing cause of *E* according to 3* because (i) *DB* is not a direct cause of *E* relative to the variable set excluding *P* (for the simple reason that *DB* and *E* are uncorrelated), and (ii) *DB* is a not direct cause of *E* when we consider richer variable sets than the original (because there is no dependence between *DB* and *E* that can be revealed by bringing to light canceling directed paths). Meanwhile, 3* still manages to say that *D1* is a contributing cause of *D3* (because *D1* is a direct cause of *D3* relative to the variable set excluding *D2*), and that *BC* is a contributing cause of *TH* (because *BC* is a direct cause of *TH* when *P* is included,

[37] 2** may seem of limited use to agents like us since we lack the capacity to contemplate entire world histories, but we can do our best. This is why the methodological advice suggested by 2** is to check whether *X* and *Y* are correlated whenever you condition on everything that *you can think of* prior to *X*. This allows the human agent to determine whether *X* is a total cause of *Y*, given her best approximation of how things might have been before *X*. The same goes for 1*, 2*, and 3*—i.e., it may seem that we're ill equipped to determine whether *X* is a cause of *Y* according to any of these conditions since it's hard to entertain causally sufficient variable sets in their entirety, but here, again, we can try our best, and can answer these queries relative to our best approximation of what's causally sufficient.

[38] Here, again, 'direct cause' should be understood in terms of 1*.

even though BC is not a direct cause of TH when P is excluded). Thus, when it comes to the examples at hand, 3* gets things exactly right.[39,40]

3*'s success at capturing the examples reflects its ability to capture the sense in which X can qualify as a cause of Y when X is a remote cause of Y and when X is not a total cause of Y without treating causal relevance (or contributing causal relevance) as the transitive closure of direct causal relevance. When there is an indirect path between X and Y (relative to some variable sets), we simply focus on a variable set that omits any causal intermediaries between X and Y in order to reveal whether X causes Y. And when X is a contributing cause of Y but not a total cause of Y (because distinct paths between X and Y cancel), we simply focus on a variable set that includes causal intermediaries along some, but not all, of the paths in order to reveal that X is a direct cause of Y along some path, and, therefore, that X is a cause of Y.

Since 3* makes no explicit reference to interventions, and can rely on 1* for its understanding of 'direct cause' in settings where causes temporally precede their effects, it is clear that we can use temporal and probabilistic information to infer when

---

39 I conjecture that 3* is extensionally equivalent to Woodward's analysis of 'contributing cause' when causes temporally precede their effects, but I have no proof. Still, this is not particularly bothersome because Woodward uses the examples under discussion to motivate his own analysis, and 3* yields the same results when applied to these cases.

40 An anonymous referee helpfully points out that there are counterexamples to 3* given some ways of formulating causal sufficiency. Consider, for example, the influential construal of causal sufficiency according to which $\mathbf{V}$ is causally sufficient exactly when there is no variable L not in $\mathbf{V}$ and two variables X, Y in $\mathbf{V}$ such that L is a direct cause of X and of Y relative to $\mathbf{V} + \{L\}$. The referee proposes the following counterexample to 3* when "causally sufficient variable set" is understood in this way.

> Imagine that in the birth control example (Fig. 2), another variable, say, "plan to have children" (PC), affects BC directly (but no other variable directly). That is, $PC \rightarrow BC \rightarrow TH$, $BC \rightarrow P \rightarrow TH$, and the two pathways between BC and TH cancel out. Then PC, intuitively, and by the interventionist's lights, is a contributing cause of TH (though not a total cause). However, the condition laid out in 3* does not seem to be satisfied. A variable set containing PC and TH either includes BC or not. If it does, then PC is not a direct cause of TH relative to the variable set; if it does not, then the variable set is not causally sufficient unless it leaves out P (and all the other intermediate variables on the path $BC \rightarrow P \rightarrow TH$) as well. But if the variable set leaves out P (and all the other intermediate variables on the path $BC \rightarrow P \rightarrow TH$), PC won't be a direct cause of TH relative to the variable set (since PC is not a total cause of TH). Therefore, there appears to be no causally sufficient set relative to which PC is a direct cause of TH.

The referee is right that this is a counterexample to 3* when "causally sufficient variable set" is understood in these terms. But I suspect that we can deal with counterexamples of this sort by revising the operative notion of causal sufficiency, rather than 3* itself. Remember that $\mathbf{V}$ should be regarded as causally sufficient when $\mathbf{V}$ does not omit any *confounding* common causes of variables included within $\mathbf{V}$. The question of how exactly we should characterize confounding common causes is substantive and correspondingly evades our grasp here. But I conjecture that common causes must be common *total* causes in order to confound, and that BC can thus appropriately be omitted from variable sets containing P and TH (since BC is not a total cause of TH). If this is right, then 3* gets the correct result that PC is a contributing cause of TH because PC is a direct cause of TH relative to the variable set that includes P but not BC. Why think that common causes must be common *total* causes in order to confound? Very roughly, when confounding takes place between X and Y, there is some correlation between X and Y that appears to reflect a causal relationship between X and Y, but that is actually fully explained by the fact that both X and Y are causally influenced by some latent common cause(s), L. In order for L's causal influence over X and Y to account for such a correlation, L must probabilistically influence X and Y along the causal paths from L to X and Y, and L cannot probabilistically influence X and Y along these paths if L is not a total cause of both X and Y. (The last sentence requires proof, but this must wait for some other occasion.).

X is a contributing cause of *Y* without considering the outcomes of any interventions, and that we can therefore determine whether *X* is causally relevant to *Y*, *simpliciter*, without considering the outcomes of any interventions. But this isn't the only upshot of 3*'s success. Because 3* makes no reference to temporal information apart from it's implicit reference to 1*, and because 1* agrees with Woodward's analysis of 'direct cause' (when it applies), 3* also shows that Woodward could have developed a far simpler analysis of 'contributing cause' according to which *X* qualifies as contributing cause of *Y* exactly when *X* is a direct cause (in Woodward's sense) of *Y* relative to some **V**. This underscores an important lesson of 3* that goes somewhat beyond the purposes of this paper. If you want to determine whether *X* causes *Y*, you will not be led astray if you posit causal relevance when you find a variable set relative to which *X* qualifies as a direct cause of *Y*, no matter whether you understand direct causal relevance in terms of 1 or 1*. Of course, you might mistakenly jump to the conclusion that *X* is *not* causally relevant to *Y*—viz., when, relative to every variable set that you have checked, *X* is not a direct cause of *Y*, but there is some variable set that you *haven't* checked relative to which *X* is a direct cause of *Y*.[41] So there can be false negatives. But there are no false positives—i.e., if you find some **V** relative to which *X* is a direct cause of *Y*, then you can safely infer that *X* is causally relevant to *Y*.

4 Causal relevance, causal decision theory, and dominance reasoning

It should be apparent by now that the analyses of causal relevance developed in the last section are useful insofar as they enable rich causal inference in the many contexts where causes temporally precede their effects, and where it is difficult to carry out interventions, or to assess the outcomes of hypothetical interventions. But if you require further convincing, or are just curious about decision-theoretic applications, then this section of the paper is for you. Here, my plan is to demonstrate that these analyses are especially useful for any decision-making agent who seeks to establish whether "dominance reasoning" is applicable.

Suppose that you are considering whether to take up smoking, and know that you would enjoy the pleasures of smoking, but are worried about the fact that people who smoke are more likely to get lung cancer than people who do not smoke. Now suppose that you live in a world that Big Tobacco once hoped was actual—namely, a world in which the correlation between smoking and lung cancer is due to a genetic condition that causally promotes smoking habits and lung cancer, and in which smoking exerts no causal influence over whether you get lung cancer. Is there any reason for you not to smoke?

Many decision theorists believe that *dominance reasoning* is applicable here because smoking doesn't cause cancer—i.e., that it's reasonable to settle on smoking for the simple reason that you prefer smoking to not smoking both in the event that

41 For example, if you never took stock of *P* when considering whether *BC* is causally relevant to *TH*, then you could mistakenly infer that *BC* is not causally relevant to *TH* (because *BC* is a direct cause of *TH* only relative to the variable set that includes *P*).

you do get lung cancer, and in the event that you don't get lung cancer.[42,43] But there are also cases where dominance reasoning is *not* applicable. Consider Bart, who hates studying, but is considering whether to study for a test because studying causally influences whether he passes (and he'd rather pass). Here, too, Bart prefers not studying to studying both if he passes, and if he fails, but it's clearly not reasonable for Bart to settle on not studying for this reason.[44]

When dominance reasoning applies, decision-making is considerably simplified for the agent. When you're deciding whether to smoke in the Big Tobacco world, you needn't think about how much you enjoy the pleasures of smoking, how much you'd hate getting lung cancer, or how probable it is that you'll get lung cancer if you smoke (or don't smoke). These difficult quantitative questions fall by the wayside since you can infer that you should smoke simply by realizing that you prefer smoking to not smoking when you get lung cancer, and when you don't. But there is no similarly streamlined way for Bart to make a decision about whether to study—i.e., he must take stock of how much he hates studying, how much he values passing, and of how probable it is that he'll pass if he studies (or doesn't study), since each of these things could plausibly tip the balance in favor of studying or not studying.[45]

Since deliberation is simplified so much by dominance reasoning, it is important to know when dominance reasoning is applicable. Causal decision theorists argue that dominance reasoning should be applied only when the possible actions (e.g., studying and not studying) are causally independent from the partition of states at hand (e.g., passing and failing). The underlying thought is that Bart can't rationally base his decision on the fact that he prefers studying no matter the outcome of the test because the outcome of the test is *causally affected* by whether Bart studies, while you *can* rationally base your decision whether to smoke on the fact that you prefer smoking no matter whether you get lung cancer because the status of your health is *not* causally affected by whether you smoke.

What exactly this conception of causal (in)dependence consists in has been a matter of controversy among decision theorists for quite some time, but interventionist methods have shed considerable light on the issue. Meek and Glymour (1994), for example, have shown that we can recover causal-decision-theoretic reasoning by treating the agent's options as interventions, and maintaining that dominance reasoning is appli-

[42] Every so-called "causal decision theorist" believes this.

[43] This kind of reasoning is called "dominance reasoning" because smoking is said to *dominate* abstaining.

[44] There are of course cases that fit this description where Bart *should* settle on passing—e.g., because Bart *really* hates studying, doesn't care much about passing, or is very confident that he'll fail even if he studies. But it's clear that Bart should not settle on not studying *for the reason* that he prefers not studying to studying both if he passes, and if he fails. Why? Because there are likewise cases that fit this description where Bart should *not* settle on not studying—e.g., when he barely minds studying, really hates failing, or is very confident that he'll pass if he studies, and otherwise won't.

[45] There may be some way of repartitioning the states (i.e., whether Bart passes or fails) such that dominance reasoning applies, but this kind of redescription is itself costly (insofar as it takes time and effort to find a partition of states relative to which dominance reasoning is applicable).

cable only when the partition of states at hand (e.g., whether you get lung cancer, or whether Bart passes) is uncorrelated with the agent's options (or interventions).[46,47]

Why does Meek and Glymour's strategy work? Recall Woodward's analysis of 'total cause' according to which X is a total cause of Y exactly when interventions on X change the probability distribution over Y. Meek and Glymour's treatment of causal independence suggests that an agent is free to deploy dominance reasoning relative to a partition of states only when the action variable, A, is not a total cause of the state variable, S. Thus, on the one hand, you can deploy dominance reasoning when deciding whether to smoke (*SM*) because *SM* is not a total cause (in Woodward's sense) of whether you get lung cancer (*LC*)—i.e., because interventions on *SM* are not correlated with *LC* (even though *SM* itself is correlated with *LC*). But on the other hand, Bart cannot deploy dominance reasoning when deciding whether to study (*ST*) because *ST* *is* a total cause of whether Bart passes or fails (*PF*)—i.e., interventions on (*ST*) *are* correlated with whether Bart passes the test (*PF*).

It is reasonable to wonder why the action variable, A, qualifies as causally independent of the state variable, S, exactly when A is not a *total* cause of S. After all, it would seem that X is causally independent of Y (in the sense that matters for decision theory) exactly when Y isn't causally influenced by X, and one of the important lessons of Woodward's interventionism is that there are cases where X is causally relevant to Y—equivalently, where X is a contributing cause of Y—in which X is not a total cause of Y.

In order to probe this issue, we should consider a case where the action variable, A, is a contributing cause of S, but isn't a total cause of S. Recall the Fig. 2 example, and suppose that you prefer not taking birth control to taking birth control no matter whether you get thrombosis. Does it follow that you should abstain from taking birth control? According to Meek and Glymour, it does. This is because interventions on *BC* are not correlated with *TH*. Is this reasonable? Arguably, yes. Even when you limit your consideration to variables within your control (by controlling for common causes), the probability that you get thrombosis given that you take birth control is the same no matter whether you take the pill (since every variable in the canceling paths is causally downstream from your action). So it seems that you should abstain from taking birth control because (i) you prefer abstaining both when you get thrombosis, and when you don't, and (ii) your abstaining from birth control does nothing to causally *promote* states of affairs in which you get thrombosis, or in which you don't.

Meek and Glymour show that we can capture causal-decision-theoretic reasoning by maintaining that dominance reasoning is applicable only when the states at hand are uncorrelated with interventions to act, but they say little to nothing about what agents should do if they are not immediately certain whether some action variable, A, is a total cause of the state variable at hand, S. This is an important question since there are many cases that we confront in real life where it is not immediately clear whether

46 In order for dominance reasoning to apply relative to a partition of states, it is of course also necessary that the agent prefers one of her options to the other(s) no matter which cell of the state partition obtains—e.g., prefers smoking to not smoking no matter whether she gets lung cancer or doesn't.

47 This follows indirectly from Meek and Glymour's more general point that causal-decision-theoretic verdicts can be recovered by maximizing conditional expected utility and treating the agent's options as interventions.

dominance reasoning is applicable—e.g., when a pregnant woman is deliberating about whether to take acetaminophen for a headache, and knows that she would prefer taking acetaminophen both in the event that her future child will be healthy, and in the event that her future child will not be healthy.[48] In these cases, we do not yet know whether the relevant interventions (e.g., intervening to take acetaminophen) are correlated with the relevant states (e.g., the future child's health).

One easy answer is that the agent should check whether her options (interventions) are correlated with the states at hand by consulting a probability distribution that is defined over the agent's options and the states.[49] But this is easier said than done. In order for the agent to discover whether the intervention on A is probabilistically independent from S without carrying out any interventions, she must have access to some joint distribution defined over both the endogenous variables in $\mathbf{V}$ as well as the exogenous intervention variable (representing her options).[50] But for the deliberating agent, it's incredibly hard to access a distribution that is defined over her options (i.e., the intervention), if these probabilities are even well defined.[51] Consider Bart's decision whether to study. He might generally know of himself that he is the kind of person who tends to study (or not study), and he might have some idea of how frequently he studies in like circumstances. But this falls short of having access to a particular probability estimate that he will decide to study as he decides whether to study. In order to entertain *these* probabilities as he decides, Bart must have some conjecture about how he'll decide as he makes up his mind about what to do. But it seems at least hard (and perhaps impossible)[52] for Bart to entertain these probabilities since the activity of making up your mind is naturally construed in terms of *determining* what you will do (or the probability that you will do something), and it is at best difficult, at worst impossible, to entertain probabilities that, by your lights, are currently in your power to determine.

It's good news, then, that one can determine whether A is a total cause of S without considering any interventions in the many settings where causes temporally precede their effects. In these settings, the agent can instead apply 2* or 2** by checking (2*) whether A is a direct cause of S (in the 1* sense) relative to some causally sufficient variable set containing S and A that excludes any variables temporally in between S and A, or (2**) whether A and S are probabilistically dependent conditional on some assignment of values over *everything* the agent can think of that is settled temporally prior to A.[53] In so doing, the agent determines whether A is a total cause of S (and,

48 It is unclear whether acetaminophen is harmful to unborn children. See Oster (2014).

49 In the same way that we use the ratio formula to check whether two variables are correlated when supplied with a frequency table (e.g., in introductory statistics classes), we might try to use the ratio formula to check whether the intervention variable is correlated with the state variable given the joint distribution.

50 I am of course assuming that the agent does not know the causal structure (because that's what she is trying to figure out).

51 Whether there are probabilities for an agent's options is another matter of great controversy among decision theorists. Since my argument goes through on the more modest claim that these probabilities are hard to access (if they exist), I have no dog in this fight. But see Levi (1997) and Spohn (1977) for classic arguments that there are no probabilities for options, and Hájek (2016) for a dissenting opinion.

52 This, again, is related to the issue of whether there are probabilities for options.

53 The 2** construal of total causal relevance may be especially useful in ordinary decision-making contexts. As you wonder whether dominance reasoning is applicable, you can simply check whether, given your

therefore, whether the probability distribution over S is affected by intervening on A) without checking the probabilities of her options (interventions), and therefore manages to check whether S is causally independent from A (in the sense that legitimates dominance reasoning) without running into any pitfalls that arise when inquiring about the probabilities of one's options.[54]

It may *prima facie* seem just as problematic (or difficult) for the agent to check whether dominance reasoning applies by checking a probability distribution that is defined over the action variable since, e.g., (i) Bart deliberates about whether to study (and thereby thinks of himself as determining whether he will study), and (ii) ST is the action variable in Bart's decision-making context. But this apparent difficulty is circumvented by the fact that, in the interventionist framework, the probabilities over endogenous variables are the probability estimates for when that variable is not intervened upon.[55] This means that Bart must only entertain the probability that he studies in the event that he does not make a decision about whether to study (since Bart's options are construed as interventions on ST by Meek and Glymour), and this seems to be the exact kind of probability that he can easily consider. It's no problem for Bart to conjecture how probable his studying would be *were he not to be deciding whether to study*, given what he knows about himself, and about what he is prone to doing in like circumstances. What's problematic, rather, is conjecturing how probable it is that he *will* study upon making the decision that he is *actually* in the process of making, and this is exactly what Bart needn't consider in order to apply 2* or 2**.

This shows that when attempting to establish whether dominance reasoning is applicable, the agent can get around the problems associated with checking whether her options (interventions) are probabilistically (in)dependent of the states at hand by using 2* or 2** to check whether the variable that she will intervene upon (by making a decision) is a total cause of the partition of states at hand. Thus there is at least one important context where the temporal analyses developed in the last section are demonstrably useful—namely, when an agent is inquiring about whether it is appropriate to apply dominance reasoning.

5 Conclusion

I now hope to have established the power of the commonsense assumption that causes temporally precede their effects from within the interventionist framework. In the great many settings where causes actually do temporally precede their effects, we can

Footnote 53 continued
best judgment, there exists some way that the past could have been conditional upon which A is correlated with S.

[54] If, for some reason, the agent finds it easier to use 3* check whether A is a contributing cause of S, this will not lead her astray (since A is a total cause of S only when A is a contributing cause of S), but she will miss out on opportunities to apply dominance reasoning when A is a contributing cause of S, but not a total cause of S.

[55] See Spirtes et al. (2000), who treat the intervention on A as a cause of A that is neither included in $\mathbf{V}$ nor a direct cause of any variable in $\mathbf{V}$ other than A, and that is partitioned such that it contains (i) a value on which one can condition to deterministically set A to a for every a in A and (ii) a value that corresponds to *not* intervening on A (i.e., to allowing the probability distribution over A to be determined by A's causes in $\mathbf{V}$).

use this assumption, together with a temporal ordering and probability distribution over the variables of interest, to infer exactly when, and in what sense, X is causally relevant to Y without considering the outcomes of any interventions. This is useful not only in the great many contexts where you want to know what causes what, but cannot intervene on the things that you care about, but also when there is some cognitive difficulty with entertaining thoughts about interventions or their outcomes—e.g., when the deliberating agent wants to know whether she can simplify the deliberative process by deploying dominance reasoning, but finds it difficult to consider the probability that she will decide in a particular way.

Acknowledgements For helpful discussion and comments, I am grateful to Benjamin Eva, Daniel Hausman, Shanna Slank, Olav Vassend, Jiji Zhang, the anonymous reviewers, and the audience at the inaugural workshop on Decision Theory and the Future of Artificial Intelligence in Cambridge, UK. This research was partially funded by Deutsche Forschungsgemeinschaft (Grant No. 623584).

References

Bramley, N., Dayan, P., Griffiths, T., & Lagnado, D. (2017). Formalizing Neurath's ship: Approximate algorithms for online causal learning. *Psychological Review, 124*, 301–338.

Burns, P., & McCormack, T. (2009). Temporal information and children's and adults' causal inferences. *Thinking and Reasoning, 15*, 167–196.

Dummett, M. (1954). Can an effect precede its cause? *Proceedings of the Aristotelian Society, 28*, 27–44.

Dummett, M. (1964). Bringing about the past. *Philosophical Review, 73*, 338–359.

Eberhardt, F. (2014). Direct causes and the trouble with soft interventions. *Erkenntnis, 79*, 755–777.

Eells, E. (1991). *Probabilistic causality*. Cambridge: Cambridge University Press.

Elwert, F., & Christakis, N. (2008). Wives and ex-wives: A new test for homogamy bias in the widowhood effect. *Demography, 45*, 851–873.

Hájek, A. (2016). Deliberation welcomes prediction. *Episteme, 13*, 507–528.

Hausman, D., & Woodward, J. (1999). Independence, invariance, and the causal Markov condition. *British Journal for the Philosophy of Science, 50*, 521–583.

Hesslow, G. (1976). Discussion: Two notes on the probabilistic approach to causality. *Philosophy of Science, 43*, 290–292.

Hitchcock, C. (2018). Causal models. In E. N. Zalta (Ed.), *The Stanford Encyclopedia of Philosophy*. https://plato.stanford.edu/entries/causal-models/. Accessed 15 Nov 2018.

Lagnado, D., & Sloman, S. (2006). Time as a guide to cause. *Journal of Experimental Psychology. Learning, Memory, and Cognition, 32*, 451–460.

Levi, I. (1997). *The covenant of reason: Rationality and the commitments of thought*. Cambridge: Cambridge University Press.

McDermott, M. (1995). Redundant causation. *The British Journal for the Philosophy of Science, 46*, 523–544.

Meek, G., & Glymour, C. (1994). Conditioning and intervening. *The British Journal for the Philosophy of Science, 45*, 1001–1021.

Oster, E. (2014). Pregnant women, here's one less thing to worry about. *FiveThirtyEight*. https://fivethirtyeight.com/features/pregnant-women-heres-one-less-thing-to-worry-about/. Accessed 15 Nov 2018.

Pearl, J. (1988). *Probabilistic reasoning in intelligent systems*. San Francisco: Morgan Kaufman.

Pearl, J. (2009). *Causality: Models, reasoning, and inference* (2nd ed.). Cambridge: Cambridge University Press.

Price, H. (1996). *Time's arrow and archimedes' point*. Oxford: Oxford University Press.

Price, H. (2001). Backward causation, hidden variables and the meaning of completeness. *Pramana, 56*, 199–209.

Spirtes, P., Glymour, C., & Scheines, R. (2000). *Causation, prediction, and search* (2nd ed.). New York: Springer.

Spohn, W. (1977). Where Luce and Krantz do really generalize savage's decision model. *Erkenntnis, 11*, 113–134.
Spohn, W. (1980). Stochastic independence, causal independence, and shieldability. *Journal of Philosophical Logic, 9*, 73–99.
Spohn, W. (2001). Bayesian nets are all there is to causal dependence. In Galavotti, Suppes, & Constantini (Eds.), *Stochastic causality* (pp. 157–172). Stanford: CSLI Publications.
Spohn, W. (2012). *The laws of belief: Ranking theory and its philosophical applications*. Oxford: Oxford University Press.
Strevens, M. (2007). Review of woodward. *Making Things Happen. Philosophy and Phenomenological Research, 74*, 233–249.
Woodward, J. (2003). *Making things happen: A theory of causal explanation*. Oxford: Oxford University Press.
Woodward, J. (2008). Response to strevens. *Philosophy and Phenomenological Research, 77*, 193–212.
Zhang, J., & Spirtes, P. (2008). Detection of unfaithfulness and robust causal inference. *Minds and Machines, 18*, 239–271.
Zhang, J., & Spirtes, P. (2011). Interventionism, determinism, and the causal minimality condition. *Synthese, 182*, 335–347.

Synthese (2021) 198 (Suppl 27):S6529–S6570
https://doi.org/10.1007/s11229-018-02010-x

Desirability foundations of robust rational decision making

Marco Zaffalon[1] · Enrique Miranda[2]

Received: 12 February 2018 / Accepted: 2 November 2018 / Published online: 11 November 2018

Abstract

Recent work has formally linked the traditional axiomatisation of incomplete preferences à la Anscombe-Aumann with the theory of desirability developed in the context of imprecise probability, by showing in particular that they are the very same theory. The equivalence has been established under the constraint that the set of possible prizes is finite. In this paper, we relax such a constraint, thus de facto creating one of the most general theories of rationality and decision making available today. We provide the theory with a sound interpretation and with basic notions, and results, for the separation of beliefs and values, and for the case of complete preferences. Moreover, we discuss the role of conglomerability for the presented theory, arguing that it should be a rationality requirement under very broad conditions.

Keywords Decision theory · Incomplete preferences · Desirability · Imprecise probability · Foundations

1 Introduction

Background

Preferences play two important roles in artificial intelligence (AI). On the one side, the axiomatisation of preferences has been a major achievement of the past century, with the pioneering work of Von Neumann and Morgestern (1947), Savage (1954) and Anscombe and Aumann (1963). These authors have justified rational decision

Chapter 7 was originally published as Zaffalon, M. & Miranda, E. Synthese (2021) 198 (Suppl 27): S6529–S6570. https://doi.org/10.1007/s11229-018-02010-x.

Work partially supported by the Swiss NSF Grant No. 200021_146606 / 2 and by Project TIN2014-59543-P.

✉ Marco Zaffalon
zaffalon@idsia.ch

Enrique Miranda
mirandaenrique@uniovi.es

1 Istituto Dalle Molle di Studi sull'Intelligenza Artificiale (IDSIA), Lugano, Switzerland

2 Department of Statistics and Operations Research, University of Oviedo, Oviedo, Spain

making through the maximisation of expected utility and at the same time have provided foundations for the Bayesian theory of probability, which is at the heart of so much work in AI. On the other side, in the recent years we have witnessed a steady increase of research in preference modelling as a ductile and practical way to assess a subject's beliefs, to learn trends from data, etc. (see, e.g., Domshlak et al. 2011; Pigozzi et al. 2016). So preferences are very flexible objects that allow us both to work out foundations for our uncertainty models and to develop useful tools for inference.

One aspect of preferences that seems particularly promising for modern artificial intelligence is their possibility to model incompleteness: the fact that we are not always capable to compare alternatives. In fact, modelling the incompleteness in our knowledge is key to develop reliable (or *robust*) applications of AI, which is a central theme now that AI is becoming so widespread. This has eminently been pointed out by Dietterich (2017), the former President of AAAI.

Research on the axiomatisation of incomplete preferences has primarily been done in communities other than AI, with the work of Bewley (2002), Seidenfeld et al. (1995), Nau (2006), Ok et al. (2012) and Galaabaatar and Karni (2013) (but see the work by Dubois et al. 2003). In fact, the theme of incompleteness or robustness is not mainstream in AI yet, even though we can find well-developed algorithms and methods, like those by Corani and Zaffalon 2008; Kikuti et al. 2011; de Bock and de Cooman 2014, just to mention a few.

What these latter proposals have in common is their reference to Walley's (1991) theory of *imprecise probability* as their founding paradigm (see Augustin et al. 2014 for a collection of introductory papers on imprecise probability and some main applications to artificial intelligence). Such a theory is a re-elaboration of Williams' (1975) theory, which, in turn, is an extension of Finetti's (1937) Bayesian theory (see also de Finetti 1974) made to deal with imprecision in probabilities, as originated by incompleteness or other reasons.

In short, Williams-Walley's theory models a subject's uncertainty by focusing on the gambles that the subject is willing to take and that for this reason are called *desirable*. The rationality of the assessments is then formalised through some axioms, which we can regard as a way to axiomatise the theory of desirability. The resulting uncertainty theory is very powerful: above all, it maintains the characteristic of logical self-consistency already present in de Finetti's theory, which it encompasses as a special case (the property of self-consistency is called *coherence* in the jargon of imprecise probability). But it is capable of modelling incompleteness in great generality too; in fact it encompasses also robust Bayesianism, and more in general the approaches that model uncertainty with sets of probabilities. And moreover it goes well beyond that, since desirability smoothly deals also with non-Archimedean problems, those that probabilities cannot capture. And it has all this done without incurring measurement problems.

Desirability and preference have essentially lived two separate lives until recently, when we have proved that they are actually the same theory (Zaffalon and Miranda 2017; an earlier report dates back to 2015). This may be surprising as preferences deal with beliefs (probabilities) and values (utilities) at the same time, while desirability was born as a theory of uncertainty alone. In fact, the equivalence has been achieved by enriching desirability models through a set of possible prizes, in addition to the usual set of states of nature.

The equivalence of the two theories thus established has proved fruitful in many respects. From the viewpoint of preference, it has allowed to show for example that the traditional notion of 'state independence', i.e., the fact that we can actually have separate models for beliefs and values, corresponds to the notion of 'strong independence' in probability (Cozman 2012): this is based on the idea of using a set of stochastically independent models. It has permitted to rework the notions of Archimedeanity in preference in a new way, and to streamline the notions of precise beliefs and values, and more in general of complete preferences.

Conversely, the equivalence has allowed us to generalise desirability so as to deal with imprecise non-linear utility. This has made of desirability one the most powerful, and well founded, theories of uncertainty and decision making available today. As such, it appears to be a natural foundation for robust methods in modern AI.

Outline and contributions

With this in mind, in this paper we revisit our work from 2017, aiming to widen its scope and to discuss more broadly some important issues that we left open. After an introduction to desirability and preferences in Sect. 2, we then give the following contributions:

> Section 3. In our 2017 formulation, we considered an unconstrained space of possibilities Ω and a finite space of prizes $\mathcal{X}$. This is particularly restrictive when we want to consider for instance continuous rewards. It is a case that has received quite some attention in the literature (see the work by del Amo and Ríos Insua 2002; Dubra et al. 2004; Ghirardato and Siniscalchi 2012; Ok et al. 2012, just to name a few) but it does not seem to have found a fully general formulation yet. In fact we also discussed this possibility pointing to some inherent difficulty.
> The main problem in this case is that the core modelling unit in preferences, that of 'horse lottery' does not easily lend itself to work with continuous sets of prizes—on top of which, by definition, we should place a probability. We address this problem by defining the new notion of *conditional horse lottery* and reformulating the axiomatisation of preferences on that basis. Then we prove the equivalence with desirability in full generality where both the spaces involved, of possibilities and prizes, are unconstrained. We make this by showing that there is a one-to-one correspondence between the axioms of preference and those of desirability.
> In the Archimedean case, this leads to represent utilities via sets of finitely additive probabilities. More generally speaking, we benefit from the representation power of desirability so as to formulate a theory that can work with any spaces, Archimedean and non-Archimedean problems, and without incurring problems of measurability. To our knowledge this type of generality is achieved here for the first time.
> Section 4 gives a rigorous interpretation of the rewards that conditional horse lotteries deliver by constructing a linear utility scale based on so-called 'probability currency'.

- Section 5 compares the traditional formalism based on horse lotteries and our new one based on conditional horse lotteries. We show that they are equivalent in the case where the prizes are in a finite number. When they are not, we discuss why our formalism appears to be more suited to deal with infinitely many prizes.
- Section 6 focuses on the question of conglomerability. Conglomerability is a property that probabilistic, as well as desirability, modelling has to face when we deal with infinite spaces of possibilities, and in particular with infinite conditioning partitions of those spaces. This is a subject of interest to us given the unconstrained nature of the space of possibilities that we consider. The main question is whether or not conglomerability should be a rationality axiom.
  Conglomerability has been first discussed by de Finetti (1931) and later on by many other authors, including Hill and Lane (1985), Armstrong (1990), Seidenfeld et al. (1998), Howson (2009), DiBella (2018). Berti et al. (2017) have recently given an overview of the subject. Despite the amount of work devoted to conglomerability, it still remains a controversial question to date.
  In this panorama, we have provided an original justification of conglomerability through arguments of 'temporal coherence', that is, the idea that an uncertainty model should be dynamically coherent in time (Zaffalon and Miranda 2013). This idea holds the promise to finally resolve the controversy, given that conglomerability turns out to be a consequence of temporal coherence. But the assumptions we imposed to derive such a result make its application relatively narrow.
  We reconsider the problem and show that we can weaken those assumptions in a substantial way, thus de facto turning conglomerability into a rationality requirement whenever a model is used to compute future beliefs and values—in the same way as probabilistic updating is usually understood.
- "Appendix A" summarises our past results about the decomposition and completeness of preferences, and proves that those results continue to hold when the space of prizes is infinite.

Our conclusions and future perspectives are in Sect. 7. The proofs of the paper have been collected in "Appendix B".

2 Preliminary notions

In this section, we introduce the main features of sets of desirable gambles and preference relations that shall be needed in the remainder of the paper.

2.1 Desirable gambles

We next introduce the basic theory of sets of desirable gambles; for a deeper account of the results and concepts mentioned in this section, we refer to the work of Walley (1991, Section 3.7); Couso and Moral (2011), de Cooman and Miranda (2012), de Cooman and Quaeghebeur (2012), Miranda (2008), Miranda and Zaffalon (2010) and Augustin et al. (2014, Chapter 1). For a philosophical defence of indeterminate probabilities, see, e.g., Hájek and Smithson (2012).

In short, the idea is to consider states of nature and associated real-valued rewards called *gambles*. The focus will be on those gambles that are 'desirable' to us and that have the property of being coherent, which means that they satisfy some rationality axioms (Definition 3 as well as Axioms D1–D4 below). A coherent set of desirable gambles determines a so-called *coherent lower prevision*, which has the interpretation of a lower expectation. In turn, that determines a set of finitely additive probabilities.[1] Coherent sets of desirable gambles, as well as coherent lower previsions and sets of probabilities, can be conditioned, marginalised or extended. The former are however more general than the latter, in that they can readily deal with conditioning on events of probability zero or with choices under zero expectation, for which the traditional, Kolmogorovian, approach to probability is not applicable.

Let us see these notions in more detail now.

2.1.1 Logical foundations of desirability

Let Ω denote the set of possible outcomes of an experiment, that is, the *space of possibilities*. We let the cardinality of Ω be general, so Ω can be infinite. A *gamble* $f : \Omega \to \mathbb{R}$ is a bounded real-valued function of Ω. A gamble is interpreted as an uncertain reward in a linear utility scale. Accepting a gamble f is regarded as a commitment to receive $f(\omega)$ whatever ω occurs.

Denote by $\mathcal{L}(\Omega)$ the set of all the gambles on Ω and by $\mathcal{L}^+(\Omega) := \{f \in \mathcal{L}(\Omega) : f \gneq 0\}$ the subset of the *positive gambles*, where $f \gneq 0$ means that $f \geq 0$ and $f \neq 0$. We denote these sets also by $\mathcal{L}$ and $\mathcal{L}^+$, respectively, when there can be no ambiguity about the space involved. Negative gambles are defined analogously by $\mathcal{L}^- := -\mathcal{L}^+$.

We examine a set of gambles $\mathcal{Q} \subseteq \mathcal{L}$ and come up with the subset $\mathcal{K}$ of the gambles in $\mathcal{Q}$ that we find desirable. How can we characterise the rationality of the assessments represented by $\mathcal{K}$?

We can follow the procedure adopted in similar cases in logic, where first of all we need to introduce a notion of deductive closure: that is, we first characterise the set of gambles that we must find desirable as a consequence of having desired $\mathcal{K}$ in the first place. This is easy to do once we assume that our utility scale is linear; those gambles are the positive linear combinations of gambles in $\mathcal{K}$:

$$\operatorname{posi}(\mathcal{K}) := \left\{ \sum_{j=1}^{r} \lambda_j f_j : f_j \in \mathcal{K}, \lambda_j > 0, r \geq 1 \right\}.$$

We must also consider that any gamble in $\mathcal{L}^+$ should be desirable as well, given that it may increase our utility without ever decreasing it. Stated differently, the set $\mathcal{L}^+$ plays the role of the tautologies in logic. This means that the actual deductive closure we are after is given by the following:

[1] Such a correspondence follows from a separating hyperplane theorem.

Definition 1 (*Natural extension for gambles*) Given a set $\mathcal{K}$ of desirable gambles, its *natural extension* $\mathcal{R}$ is the set of gambles given by

$$\mathcal{R} := \text{posi}(\mathcal{K} \cup \mathcal{L}^+).$$

Note that $\mathcal{R}$ is the smallest convex cone that includes $\mathcal{K} \cup \mathcal{L}^+$.

The rationality of the assessments is characterised through the natural extension by the following:

Definition 2 (*Avoiding partial loss for gambles*) A set $\mathcal{K}$ of desirable gambles is said to *avoid partial loss* if $0 \notin \mathcal{R}$.

This condition is the analog of the notion of consistency in logic. The irrationality of a natural extension that incurs partial loss depends on the fact (as it is possible to show) that it must contain a negative gamble f, that is, one that cannot increase our utility and can possibly decrease it. In contradistinction, a set that avoids partial loss does not contain negative gambles.

There is a final notion that is required to make a full logical theory of desirability. This is the logical notion of a theory, that is, a set of assessments that is consistent and logically closed, in the sense that the consistent assessments coincide with their deductive closure in the examined domain $\mathcal{Q}$. This means that we are fully aware of the implications of our assessments on other elements of $\mathcal{Q}$. The logical notion of a theory goes in desirability under the name of coherence:

Definition 3 (*Coherence for gambles*) Say that $\mathcal{K}$ is *coherent relative to* $\mathcal{Q}$ if $\mathcal{K}$ avoids partial loss and $\mathcal{Q} \cap \mathcal{R} \subseteq \mathcal{K}$ (and hence $\mathcal{Q} \cap \mathcal{R} = \mathcal{K}$). In case $\mathcal{Q} = \mathcal{L}$ then we simply say that $\mathcal{K}$ is *coherent*.

This definition alone, despite its conceptual simplicity, makes up all the theory of desirable gambles: in principle, every property of the theory can be derived from it. Moreover, the definition gives the theory a solid logical basis and in particular guarantees that the inferences one draws are always coherent with one another. This appears to giving a reply to Howson's (2009) question, whether probability and logic can be combined (see also Howson 2011): by reformulating 'probability' as desirability, we do not even need to combine probability with logic, as we clearly see that probability is already pure logic. At the same time, reformulating probability through desirability, we obtain a very powerful theory: as we have seen, it can be defined on any space of possibilities Ω and any domain $\mathcal{Q} \subseteq \mathcal{L}$ (in this sense, it is not affected by measurability problems); and, as we shall make precise later on in this section, it can handle both precise and imprecise assessments, as well as model both Archimedean and non-Archimedean problems.

2.1.2 The case $\mathcal{Q} = \mathcal{L}$

In this paper we shall primarily be concerned with the case where the set of gambles $\mathcal{Q}$ that we examine is the entire set $\mathcal{L}$. We shall then focus on this case to develop the theory of desirability in some detail.

The first consequence of focusing on the case $\mathcal{Q} = \mathcal{L}$ is that coherence can equivalently be characterised by four axioms (see Walley 1991, Section 3.7; Miranda and Zaffalon 2010, Proposition 2):

Definition 4 (*Coherence for gambles when* $\mathcal{Q} = \mathcal{L}$) In the case where $\mathcal{Q} = \mathcal{L}$, $\mathcal{R}$ is said to be a *coherent set of desirable gambles* in $\mathcal{L}$ if and only if it satisfies the following conditions:

D1. $\mathcal{L}^+ \subseteq \mathcal{R}$ [Accepting Partial Gains];
D2. $0 \notin \mathcal{R}$ [Avoiding Null Gain];
D3. $f, g \in \mathcal{R} \Rightarrow f + g \in \mathcal{R}$ [Additivity];
D4. $f \in \mathcal{R}, \lambda > 0 \Rightarrow \lambda f \in \mathcal{R}$ [Positive Homogeneity].

A coherent set of desirable gambles is thus a convex cone (D3, D4) that excludes the origin (D2) and that contains the positive gambles (D1).

Consider an event $B \subseteq \Omega$. We denote by B both the subset of Ω and its indicator function I_B (that equals one in B and zero elsewhere). Using this convention, we can multiply B and a gamble f obtaining the gamble Bf given for all $\omega \in \Omega$ by

$$Bf(\omega) = \begin{cases} f(\omega) & \text{if } \omega \in B, \\ 0 & \text{otherwise.} \end{cases}$$

Definition 5 (*Conditioning for gambles*) Consider a coherent set of desirable gambles $\mathcal{R}$ on $\mathcal{L}$ and let B be a non-empty subset of Ω. The set of desirable gambles conditional on B derived from $\mathcal{R}$ is defined as

$$\mathcal{R}|B := \{f \in \mathcal{R} : f = Bf\}.$$

There is a natural correspondence between Bf and the restriction of f to B, whence we can also put $\mathcal{R}|B$ in relation with

$$\mathcal{R}\rfloor B := \{f_B \in \mathcal{L}(B) : (\exists f \in \mathcal{R}|B)(\forall \omega \in B)\ f(\omega) = f_B(\omega)\},$$

which is a coherent set of desirable gambles in $\mathcal{L}(B)$. $\mathcal{R}|B$ and $\mathcal{R}\rfloor B$ are equivalent representations of the conditional set.

Definition 6 (*Marginalisation for gambles*) Let $\mathcal{R}$ be a coherent set of desirable gambles in the product space $\Omega \times \Omega'$, where Ω, Ω' are two logically independent sets. The *Ω-marginal set of desirable gambles on $\mathcal{L}(\Omega \times \Omega')$ induced by $\mathcal{R}$* is defined as

$$\mathcal{R}_\Omega := \{f \in \mathcal{R} : (\forall \omega \in \Omega)(\forall \omega'_1, \omega'_2 \in \Omega')\ f(\omega, \omega'_1) = f(\omega, \omega'_2)\}.$$

Since $\mathcal{R}_\Omega$ is made up of gambles that depend on Ω only (we call them *Ω-measurable* gambles), we can establish a bijection between $\mathcal{R}_\Omega$ and

$$\mathcal{R}'_\Omega := \{g \in \mathcal{L}(\Omega) : (\exists f \in \mathcal{R})(\forall \omega \in \Omega)(\forall \omega' \in \Omega')\ g(\omega) = f(\omega, \omega')\}.$$

$\mathcal{R}'_\Omega$ is a coherent set of gambles in $\mathcal{L}(\Omega)$.

A coherent set of desirable gambles encompasses a probabilistic model for Ω made of lower and upper expectations (also called *previsions* after de Finetti):

Definition 7 (*Coherent lower and upper previsions*) Let $\mathcal{R}$ be a coherent set of desirable gambles in $\mathcal{L}$. For all $f \in \mathcal{L}$, let

$$\underline{P}(f) := \sup\{\mu \in \mathbb{R} : f - \mu \in \mathcal{R}\}, \tag{1}$$

where addition of a gamble with a constant is meant pointwise. It is called the *lower prevision* of f. The conjugate value given by $\overline{P}(f) := -\underline{P}(-f)$ is called the *upper prevision* of f. The functionals $\underline{P}, \overline{P} : \mathcal{L} \to \mathbb{R}$ are respectively called a *coherent lower prevision* and a *coherent upper prevision*.

Definition 8 (*Linear prevision*) Let $\underline{P}, \overline{P}$ be coherent lower and upper previsions on $\mathcal{L}$. If $\underline{P}(f) = \overline{P}(f)$ for some $f \in \mathcal{L}$, then we call the common value the *prevision* of f and we denote it by $P(f)$. If this happens for all $f \in \mathcal{L}$ then we call the functional P a *linear prevision*.

A coherent lower prevision $\underline{P}$ has a set of *dominating* linear previsions:

$$\mathcal{M}(\underline{P}) := \{P \text{ linear prevision} : (\forall f \in \mathcal{L})\ P(f) \geq \underline{P}(f)\}, \tag{2}$$

which turns out to be closed[2] and convex. Since each linear prevision is in a one-to-one correspondence with a finitely additive probability, we can regard $\mathcal{M}(\underline{P})$ also as a set of probabilities (or a *credal set*). Moreover, $\underline{P}$ is the lower envelope of the previsions in $\mathcal{M}(\underline{P})$:

$$(\forall f \in \mathcal{L})\ \underline{P}(f) = \min\{P(f) : P \in \mathcal{M}(\underline{P})\}.$$

The coherent upper prevision $\overline{P}$ is the upper envelope of the same set; as a consequence, it follows that $\underline{P}(f) \leq \overline{P}(f)$ for all $f \in \mathcal{L}$.

Definition 9 (*Strict desirability*) A coherent set of gambles $\mathcal{R}$ is said to be *strictly desirable* if it satisfies the following condition:

D0. $f \in \mathcal{R}\backslash\mathcal{L}^+ \Rightarrow (\exists\delta > 0)\ f - \delta \in \mathcal{R}$ [Openness].

Strict desirability means that $\mathcal{R}\backslash\mathcal{L}^+$ does not include its topological border. By an abuse of terminology, $\mathcal{R}$ is said to be open too.[3]

There is a one-to-one correspondence between coherent lower previsions and strictly desirable sets: from $\underline{P}$ we can induce set

$$\mathcal{R} := \{f \in \mathcal{L} : f \gneq 0 \text{ or } \underline{P}(f) > 0\}; \tag{3}$$

[2] In the weak* topology, which is the smallest topology for which all the evaluation functionals given by $f(P) := P(f)$, where $f \in \mathcal{L}$, are continuous.

[3] A note of caution to prevent confusion in the reader: the adjective 'strict' denotes two unrelated things in desirability and in preferences. In preferences it characterises irreflexive relations, while in desirability it formalises an Archimedean condition as it will become clear in Sect. 3. We are keeping the same adjective in both cases for historical reasons and given that there should be no possibility to create ambiguity by doing so.

$\mathcal{R}$ is coherent and strictly desirable and moreover induces $\underline{P}$ through Eq. (1).

Remark 1 (*General vs strict and almost desirability*) Note that a general coherent set of desirable gambles is a cone that need not be open nor closed. The closure of a coherent set of desirable gambles is called a set of *almost-desirable* gambles by Walley (1991, Section 3.7.3). Strict desirability, as we have seen, is associated with open cones. Both almost and strictly desirable gambles are in a one-to-one correspondence with, and are therefore as informative as, a closed convex set of linear previsions. The extra generality of sets of desirable gambles makes them useful when dealing with the problem of conditioning on sets of probability zero, or of choosing between two options under zero expectation (Zaffalon and Miranda 2013, Section 5; see also Dubins 1975; Seidenfeld et al. 1995; Coletti and Scozzafava 2002; de Bock and de Cooman 2015a for other approaches to this problem).

Remark 2 (*On the regularity assumption*) Note also that with coherent sets of desirable gambles we need not enter the controversy as to whether or not we should use the 'regularity assumption', which prescribes that probabilities of possible events should be positive (see Pedersen 2014). Such an assumption goes back to an important article by Shimony (1955). In the language of this paper, Shimony argued that de Finetti's framework could lead to the questionable undesirability of a positive gamble in case zero probabilities were present, given that the prevision (expectation) of such a gamble could be zero. In other words, Shimony argued, as we do, that D1 is an axiom of rationality. This led Shimony and a number of later authors, among whom Carnap (1980) and Skyrms (1980), to advocate strengthening de Finetti's theory by requiring regularity. But it has originated also much controversy given the very constraining nature of regularity on probabilistic models. Between requiring regularity and dropping the desirability of positive gambles, coherent sets of desirable gambles actually offer a third possible way that keeps all desiderata together: events can have zero probability, positive gambles are desirable, and all operations, such as conditioning or comparisons, are well defined and informative. ♢

Definition 10 (*Conditional coherent lower and upper previsions*) Let $\mathcal{R}$ be a coherent set of desirable gambles in $\mathcal{L}(\Omega)$ and B a non-empty subset of Ω. For all $f \in \mathcal{L}(\Omega)$, let

$$\underline{P}(f|B) := \sup\{\mu \in \mathbb{R} : B(f - \mu) \in \mathcal{R}\} \tag{4}$$

be the *conditional lower prevision* of f given B. The conjugate value given by $\overline{P}(f|B) := -\underline{P}(-f|B)$ is called the *conditional upper prevision* of f. The functionals $\underline{P}(\cdot|B), \overline{P}(\cdot|B) : \mathcal{L}(\Omega) \to \mathbb{R}$ are respectively called a *conditional coherent lower prevision* and a *conditional coherent upper prevision*.

Denote by $\inf_B f$ the infimum value that f takes on B. $\underline{P}(\cdot|B)$ satisfies the following conditions for all $f \in \mathcal{L}(\Omega)$ and all real $\lambda > 0$:

- $\underline{P}(f|B) \geq \inf_B f$;
- $\underline{P}(\lambda f|B) = \lambda \underline{P}(f|B)$;
- $\underline{P}(f + g|B) \geq \underline{P}(f|B) + \underline{P}(g|B)$.

These conditions are sometimes used as axioms of coherent conditional lower previsions, when not taking desirability as the primitive concept. Note that coherent lower previsions can be defined as a special case of conditional ones obtained when $B = \Omega$.

Definition 11 (*Maximal coherent set of gambles*) Let $\mathcal{R}$ be a coherent set of desirable gambles. It is called *maximal* if

$$(\forall f \in \mathcal{L}\setminus\{0\})\ f \notin \mathcal{R} \Rightarrow -f \in \mathcal{R}.$$

Requiring maximality is tantamount to assuming complete preferences.

It is important for this paper to also say something about *conglomerability*:

Definition 12 (*Conglomerability*) Consider a coherent set of desirable gambles $\mathcal{R} \subseteq \mathcal{L}(\Omega)$ and let $\mathcal{B}$ be a partition of Ω. $\mathcal{R}$ is said to be *$\mathcal{B}$-conglomerable* if it satisfies the following condition:[4]

$$(\forall f \in \mathcal{L}(\Omega))((\forall B \in \mathcal{B})\ Bf \in \mathcal{R} \cup \{0\} \Rightarrow f \in \mathcal{R} \cup \{0\}).$$

The rationale behind conglomerability is that if a gamble is desirable conditional on each event of a partition of Ω, then it should be unconditionally desirable. We refer to the work by Miranda et al. (2012) and Walley (1991, Chapter 6) for a discussion of conglomerability in terms of sets of desirable gambles. See also Sect. 6.

Definition 13 (*Marginal extension for gambles*) Let $\mathcal{R}_\Omega$ be a marginal coherent set of gambles in $\mathcal{L}(\Omega \times \Omega')$ and $\mathcal{R}|\{\omega\}$, for all $\omega \in \Omega$, be conditional coherent sets. Let

$$\mathcal{R}|\Omega := \{h \in \mathcal{L}(\Omega \times \Omega') : (\forall \omega \in \Omega)\ h(\omega, \cdot) \in \mathcal{R}|\{\omega\} \cup \{0\}\}\setminus\{0\}$$

be a set that conglomerates all the conditional information along the partition $\{\{\omega\} \times \Omega' : \omega \in \Omega\}$ of $\Omega \times \Omega'$ (by an abuse of notation, we denote this partition by Ω too). Then the following coherent and Ω-conglomerable set of gambles:

$$\hat{\mathcal{R}} := \{g + h : g \in \mathcal{R}_\Omega \cup \{0\}, h \in \mathcal{R}|\Omega \cup \{0\}\}\setminus\{0\},$$

is called the *marginal extension* of the given marginal and conditional information.

The marginal extension is a generalisation of the law of total expectation to desirable gambles. It can be defined for lower previsions too (see for instance Walley 1991, Section 6.7 and also Miranda and de Cooman 2007). To this end, we first give some preliminary notions.

Definition 14 (*Separately coherent conditional lower prevision*) Let $\mathcal{B}$ be a partition of Ω and $\underline{P}(\cdot|B)$ a coherent lower prevision conditional on B for all $B \in \mathcal{B}$. Then we call

$$\underline{P}(\cdot|\mathcal{B}) := \sum_{B \in \mathcal{B}} B\underline{P}(\cdot|B)$$

a *separately coherent conditional lower prevision*.

[4] In the following we shall just call it conglomerability for short.

For every gamble f, $\underline{P}(f|\mathcal{B})$ is the gamble on Ω that equals $\underline{P}(f|B)$ for $\omega \in B$; so it is a $\mathcal{B}$-measurable gamble.

Definition 15 (*Marginal coherent lower prevision*) Let $\underline{P}$ be a coherent lower prevision on $\mathcal{L}(\Omega \times \Omega')$. Then the *$\Omega$-marginal coherent lower prevision* it induces is given by

$$\underline{P}_\Omega(f) := \underline{P}(f)$$

for all $f \in \mathcal{L}(\Omega \times \Omega')$ that are Ω-measurable.

The Ω-marginal is simply the restriction of $\underline{P}$ to the subset of gambles in $\mathcal{L}(\Omega \times \Omega')$ that only depend on elements of Ω. For this reason, and analogously to the case of desirability, we can represent the Ω-marginal in an equivalent way also through the corresponding lower prevision $\underline{P}'_\Omega$ defined on $\mathcal{L}(\Omega)$.

Definition 16 (*Marginal extension for lower previsions*) Consider the possibility space $\Omega \times \Omega'$ and its partition $\{\{\omega\} \times \Omega' : \omega \in \Omega\}$. We shall denote this partition by Ω and its elements by $\{\omega\}$, with an abuse of notation. Let $\underline{P}'_\Omega$ be a marginal coherent lower prevision and let $\underline{P}(\cdot|\Omega)$ be a separately coherent conditional lower prevision on $\mathcal{L}(\Omega \times \Omega')$. Then the *marginal extension of $\underline{P}'_\Omega$ and $\underline{P}(\cdot|\Omega)$* is the lower prevision $\underline{P}$ given for all $f \in \mathcal{L}(\Omega \times \Omega')$ by

$$\underline{P}(f) := \underline{P}'_\Omega(\underline{P}(f|\Omega)).$$

2.2 Preference relations

We turn now our attention to the second pillar of this paper: preference relations.

Let Ω denote, as before, the space of possibilities. In order to deal with preferences, we introduce now another set $\mathcal{X}$ of outcomes, or prizes. We assume that all the pairs of elements in $\Omega \times \mathcal{X}$ are possible or, which is equivalent, that Ω and $\mathcal{X}$ are logically independent.

The treatment of preferences in this paper relies on the notion of a conditional horse lottery:

Definition 17 (*Conditional horse lottery*) A *conditional horse lottery* is a function $p : \Omega \times \mathcal{X} \to [0, 1]$.

We should recall that the modelling of preferences is traditionally done by *horse lotteries* since the early work of Anscombe and Aumann (1963): those are similar to our conditional horse lotteries but actually correspond to placing a probability over prizes for each $\omega \in \Omega$ rather than an unnormalised function, as in Definition 17. Prominent examples of axiomatisations of incomplete preferences through horse lotteries are the works of Bewley (2002), Seidenfeld et al. (1995), Nau (2006), Galaabaatar and Karni (2013) and Zaffalon and Miranda (2017). Since it is not immediate to work with horse lotteries in the case of infinitely many prizes, in this paper we have decided to define conditional horse lotteries. The essence of the two approaches is the same but technically the latter appears more naturally suited for the extension to the general case. We shall analyse the differences of the two approaches in detail in Sect. 5.

Let us denote by $\mathcal{H}(\Omega \times \mathcal{X}) := [0,1]^{\Omega \times \mathcal{X}}$ the set of all conditional horse lotteries on $\Omega \times \mathcal{X}$. They will also be called *acts* for short. In the following we shall use the notation $\mathcal{H}$ for the set of all the acts in case there is no possibility of ambiguity.

Conditional horse lotteries reward us with so-called 'probability currency'. In particular, if event ω occurs, conditional horse lottery p rewards us with an increase of probability to win prize x, for all $x \in \mathcal{X}$, proportional to $p(\omega, x)$. The reward process will be detailed in Sect. 4.[5]

Now it is convenient to give a name to the special act that provides us with no actual reward, no matter the ω that eventually occurs:

Definition 18 (*Zero act*) Let $0 \in \mathcal{H}$ denote the zero act, which is defined by $0(\omega, x) := 0$ for all $\omega \in \Omega, x \in \mathcal{X}$.

Conditional horse lotteries are related to a behavioural interpretation through a notion of preference. The idea is that, since we aim at receiving a prize in $\mathcal{X}$, we will prefer some acts over some others; this will depend in part on our knowledge about the experiment originating an $\omega \in \Omega$, and in part on our attitude, or liking, towards the elements of $\mathcal{X}$. We are particularly interested in some rational type of preference relations that we call coherent:

Definition 19 (*Coherent preference relation*) A *preference relation* $\succ$ over conditional horse lotteries is a subset of $\mathcal{H} \times \mathcal{H}$. It is said to be *coherent* if it satisfies the next four axioms:

A1. $(\forall p \in \mathcal{H}\backslash\{0\})\ p \succ 0$ [Worst act];
A2. $(\forall p \in \mathcal{H})\ p \not\succ p$ [Irreflexivity];
A3. $(\forall p, q, r \in \mathcal{H})\ p \succ q \succ r \Rightarrow p \succ r$ [Transitivity];
A4. $(\forall p, q, r \in \mathcal{H})\ p \succ q \Leftrightarrow (\forall \alpha \in (0,1])\ \alpha p + (1-\alpha) r \succ \alpha q + (1-\alpha) r$ [Mixture Independence].

If also the next axiom is satisfied, then we say that the coherent preference relation is *weakly Archimedean*.

A0. $(\forall p, q \in \mathcal{H} : \neg(p \gtrsim q))\ p \succ q \Rightarrow (\exists \alpha \in (0,1))\ \alpha p \succ q$ [Weak Archimedeanity].

Let us comment on the given axioms.

- Axiom A1 is a direct consequence of the implicit assumption that the elements of $\mathcal{X}$ are prizes, that is, objects we would like to get; since the zero act surely prevents us from increasing our chances to win them—as opposed to any other act—, then it must be the worst possible act for us.
- Axioms A2 characterises the property of being 'strict' of the preferences we are defining. In particular, by writing $p \succ q$ we want to express that p is strictly more valuable than q for us. For this reason, there cannot be p such that $p \succ p$.

[5] Our conditional horse lotteries should not be confused with Luce and Krantz's (1971) conditional acts. In such a case acts are conditional on subsets of Ω and are defined with the aim to increase their theory's expressiveness. Our horse lotteries p are conditional in the sense that $p(\omega, \cdot)$ intuitively represents a collection of conditional probabilities. This is just a technical point that makes our theory simpler and easier to extend to the case of infinitely many prizes compared to the case of horse lotteries.

- As for Axiom A3, the expression $p \succ q \succ r$ means that for us p is strictly more valuable than q, which is strictly more valuable than r; if we take this to mean that we would pay a positive amount to exchange r for q, and another to exchange q for p, then we would actually pay a positive amount to exchange r for p.
- Axiom A4 can be interpreted in different ways. One way to look at it, is that what matters for establishing that $p \succ q$ is the increase, or decrease, of probabilities that comes along with dropping act q for p, and that can be expressed by the element-wise difference $p - q$. Axiom A4 will be discussed more widely in Sect. 5.
- Finally, Axiom A0 expresses a continuity property of the model. This is easier to see by considering a violation of the axiom, where there is a preference $p \succ q$ such that any possible worsening of p would invalidate the preference itself: that is, $\alpha p \not\succ q$ for all $\alpha \in (0, 1)$. In this case there is not continuity in the transition from $\alpha p \not\succ q$ for all $\alpha \in (0, 1)$, to $p \succ q$.
  Note that the axiom excludes from consideration the cases where $p \gneq q$. The reason is that requiring continuity in those cases would conflict with rationality. Imagine for instance the case where our preference relation is given, for all $r, s \in \mathcal{H}$, by $r \succ s$ if and only if $r \gneq s$. Consider acts p, q given by $p(\omega, x) := q(\omega, x) := \frac{1}{2}$ for all $(\omega, x) \in \Omega \times \mathcal{X}$ except for the pair (ω', x'), for which we have $p(\omega', x') > q(\omega', x')$. In this situation there is no $\alpha \in (0, 1)$ such that $\alpha p \succ q$. And yet, of course, we are fully rational in expressing $p \succ q$. Therefore continuity of preferences does not seem justified when it comes to preferences that arise from dominances like $p \gneq q$. See also Proposition 2 below.

A simple yet convenient consequence of the axioms is the following:

Proposition 1 *Suppose that for given $p, q, r, s \in \mathcal{H}$ it holds that $\alpha(p - q) = (1 - \alpha)(r - s)$ for some $\alpha \in (0, 1)$. Then for any preference relation $\succ$ that satisfies mixture independence, it holds that*

$$p \succ q \Leftrightarrow r \succ s.$$

Let us consider now a special type of preference.

Definition 20 (*Objective preference*) Given acts $p, q \in \mathcal{H}$, we say that *p is objectively preferred to q* if $p \geq q$ and $p \neq q$ (we denote objective preference also by $p \gneq q$).

The idea is that p is objectively preferred to q because the probability p assigns to outcomes is always not smaller than that of q while being strictly greater somewhere. An objective preference is indeed a preference:

Proposition 2 *Let $\succ$ be a coherent preference relation on $\mathcal{H} \times \mathcal{H}$ and $p, q \in \mathcal{H}$. Then $p \gneq q \Rightarrow p \succ q$.*

This proposition remarks in a formal way that objective preferences are those every rational subject expresses. These preferences are therefore always belonging to any coherent preference relation. If they are *the only* preferences in the relation, then we call the relation *vacuous* because we are expressing only a non-informative, or trivial, type of preferences.

3 Equivalence of desirability and preference

In the generalisation of desirability we are about to describe in this section, $\mathcal{L}$ will be redefined to be the set of gambles on $\Omega \times \mathcal{X}$. These may be considered as functional-valued gambles: for each $\omega \in \Omega$, gamble $f \in \mathcal{L}(\Omega \times \mathcal{X})$ delivers the reward function $f(\omega, \cdot)$ whose elements are measured in 'probabilistic' units associated with the respective prizes (see Sect. 4 for more details). $\mathcal{R}$ henceforth will denote a coherent set of desirable gambles of this form, which is a subset of the new $\mathcal{L}$. Such a generalised definition of gambles makes them very close to conditional horse lotteries.

On this basis, we shall first show that each axiom of desirability has an equivalent counterpart among the axioms of preference. From that, we will prove the equivalence of the two theories.

3.1 Axiom-to-axiom equivalence

In this section we shall start work with preferences that are not necessarily coherent, and we shall discuss then their axioms, in relations to those of desirability, one by one. To this end, we shall initially consider only a minimal requirement on preference relations:

Definition 21 (*Linear preference relation*) We say that a relation $\succ \subseteq \mathcal{H} \times \mathcal{H}$ is *linear* if $p \succ q \Rightarrow r \succ s$ for all r, s such that $p - q = r - s$.

Now we consider a special case of sets of desirable gambles that are in a bijective relation with preferences.

Lemma 1 *Let* $\mathcal{L}_1(\Omega \times \mathcal{X}) := \{f \in \mathcal{L}(\Omega \times \mathcal{X}) : \sup |f| \leq 1\}$ *and* $\mathcal{L}_1^+ := \mathcal{L}_1 \cap \mathcal{L}^+$. *Sets of desirable gambles in* $\mathcal{L}_1(\Omega \times \mathcal{X})$ *and linear preference relations in* $\mathcal{H} \times \mathcal{H}$ *are in a one-to-one correspondence, where given a set of desirable gambles* $\mathcal{D}$, *we consider the preference relation* $\succ$ *as* $p \succ q \Leftrightarrow p - q \in \mathcal{D}$ *and given* $\succ$ *we define* $\mathcal{D} := \{f \in \mathcal{L}_1(\Omega \times \mathcal{X}) : (\exists p, q \in \mathcal{H})\ f = p - q, p \succ q\}$.

Let us define coherence axioms for the special class of desirable gambles in $\mathcal{L}_1$.

Definition 22 (*Desirability axioms for linear relations*) Consider a set of gambles $\mathcal{D} \subseteq \mathcal{L}_1$. We say that:

- *D0′*. $\mathcal{D}$ is strictly desirable if only if for all $f \in \mathcal{D} \backslash \mathcal{L}_1^+$ and all $g \in \mathcal{L}_1$ such that $\max(0, -f) \leq g \leq \min(1, 1-f)$, there is $\alpha \in (0, 1)$ such that $\alpha f - (1-\alpha) g \in \mathcal{D}$.
- *D1′*. $\mathcal{D}$ accepts partial gains if and only if $\mathcal{L}_1^+ \subseteq \mathcal{D}$.
- *D2′*. $\mathcal{D}$ avoids null gain if and only if $0 \notin \mathcal{D}$.
- *D3′*. $\mathcal{D}$ is additive if and only if for all $f, g \in \mathcal{D}$ such that $f = p - q$ and $g = q - r$ for some $p, q, r \in \mathcal{H}$, it holds that $f + g \in \mathcal{D}$.
- *D4′*. $\mathcal{D}$ satisfies positive homogeneity if and only if $f \in \mathcal{D}$, $\lambda > 0$ such that $\lambda f \in \mathcal{L}_1$, implies that $\lambda f \in \mathcal{D}$.

Each of these axioms is in a one-to-one correspondence with one of the axioms of preferences:

Theorem 1 *Let $\mathcal{D}$ and $\succ$ be respectively a set of gambles in $\mathcal{L}_1$ and the corresponding linear relation. Then:*

(A1 $\Leftrightarrow$ D1′) Relation $\succ$ has worst outcome 0 if and only if $\mathcal{D}$ accepts partial gains.
(A2 $\Leftrightarrow$ D2′) Relation $\succ$ is irreflexive if and only if $\mathcal{D}$ avoids null gain.
(A3 $\Leftrightarrow$ D3′) Relation $\succ$ is transitive if and only if $\mathcal{D}$ is additive.
(A4 $\Leftrightarrow$ D4′) Relation $\succ$ satisfies mixture independence if and only if $\mathcal{D}$ satisfies positive homogeneity.
(A0 $\Leftrightarrow$ D0′) Relation $\succ$ is weakly Archimedean if and only if $\mathcal{D}$ is strictly desirable.

3.2 Equivalence of the theories

At this point we start considering the case where the basic rationality axioms hold jointly.

Lemma 2 *Let $\mathcal{D}$ be a set of gambles in $\mathcal{L}_1$. If $\mathcal{D}$ satisfies D4′, then D3′ is equivalent to the following:*

D3″. For all $f, g \in \mathcal{D}$ such that $f + g \in \mathcal{L}_1$, it holds that $f + g \in \mathcal{D}$.

Recall that in the previous section we have shown that there is a one-to-one correspondence between linear preference relations and sets of desirable gambles in $\mathcal{L}_1$. Now we want to show what is the one-to-one correspondence between *coherent* preference relations and *coherent* sets of desirable gambles (whether in $\mathcal{L}_1$ or $\mathcal{L}$). To this end, we associate with a set of gambles $\mathcal{D} \subseteq \mathcal{L}_1$ the set

$$\mathcal{R}_{\mathcal{D}} := \{\lambda g : \lambda > 0, g \in \mathcal{D}\} \subseteq \mathcal{L}. \tag{5}$$

Vice versa, with each set of desirable gambles $\mathcal{R} \subseteq \mathcal{L}$, we associate the set

$$\mathcal{D}_{\mathcal{R}} := \left\{\frac{f}{\sup|f|} : 0 \neq f \in \mathcal{R}\right\} \subseteq \mathcal{L}_1. \tag{6}$$

Theorem 2 1. *Let $\mathcal{D}$ be a set of gambles in $\mathcal{L}_1$. If $\mathcal{D}$ satisfies D1′–D4′, then $\mathcal{R}_{\mathcal{D}}$ is coherent.*
2. *Conversely, given a coherent set of desirable gambles $\mathcal{R} \subseteq \mathcal{L}$ the set $\mathcal{D}_{\mathcal{R}}$ satisfies D1′–D4′.*
3. *Under coherence of $\mathcal{R}$ and $\mathcal{D}$, the procedures in the two statements above are each other's inverses: $\mathcal{R}_{\mathcal{D}_{\mathcal{R}}} = \mathcal{R}$ and $\mathcal{D}_{\mathcal{R}_{\mathcal{D}}} = \mathcal{D}$.*
4. *If $\mathcal{D}$ satisfies D1′–D4′, then $\mathcal{D}$ is coherent relative to $\mathcal{L}_1$, and $\mathcal{R}_{\mathcal{D}}$ is its natural extension.*
5. *Under coherence, $\mathcal{R}$ is a set of strictly desirable gambles if and only if $\mathcal{D}_{\mathcal{R}}$ satisfies D0′.*

Among other things, this theorem states that coherent preference relations are in a one-to-one correspondence with coherent sets of desirable gambles in $\mathcal{L}_1$, whose form is given in (6) when $\mathcal{R}$ is coherent. They are obviously also in a one-to-one correspondence with coherent sets of desirable gambles in $\mathcal{L}$ through (5).

We summarise the situation in the next theorem:

Theorem 3 *There is a one-to-one correspondence between coherent preference relations on $\mathcal{H} \times \mathcal{H}$ and coherent sets of desirable gambles in $\mathcal{L}(\Omega \times \mathcal{X})$. Moreover, strict desirability is in a one-to-one correspondence with weak Archimedeanity.*

3.3 Other axioms considered in the literature

Our previous results show that there is a correspondence between coherent preference relations and coherent sets of desirable gambles. In fact, this correspondence can be extended beyond the basic axioms of coherence: other interesting rationality conditions that can be imposed onto a preference relation can be formulated in terms of desirable gambles. Let us briefly discuss how this can be done.

One such axiom is called *negative transitivity*, and can be expressed as

A5. $(\forall p, q, r \in \mathcal{H})\ p \not\succ q$ and $q \not\succ r \Rightarrow p \not\succ r$ [Negative transitivity].

Negative transitivity on constant acts has been considered by Galaabaatar and Karni (2013), who related it to Knightian uncertainty. In terms of sets of desirable gambles, we can prove the following:

Proposition 3 *Let $\succ$ be a coherent preference relation on $\mathcal{H} \times \mathcal{H}$, and let $\mathcal{R}$ be its associated coherent set of desirable gambles in $\mathcal{L}(\Omega \times \mathcal{X})$. Then $\succ$ is negatively transitive if and only if $\mathcal{R}$ satisfies*

$$f, g \notin \mathcal{R} \Rightarrow f + g \notin \mathcal{R}. \tag{7}$$

Interestingly, if a coherent set of desirable gambles satisfies Eq. (7) then not only $\text{posi}(\mathcal{R}) = \mathcal{R}$ (due to coherence), but also $\text{posi}(\mathcal{R}^c) = \mathcal{R}^c$ (due to (7)), and as a consequence it corresponds to what has been called a *lexicographic set of desirable gambles* by Van Camp et al. (2018b). The reason is that these sets of gambles are in a correspondence with lexicographic probabilities. See also "Appendix A.2.3" for a somewhat related study of condition (7); in the framework of coherent lower previsions, we refer to the work of Miranda and Zaffalon (2016).

Finally, another relevant axiom in the literature is the dominance axiom:

A6. $(\forall \omega \in \Omega)\ p \succ q^{\omega} \Rightarrow p \succ q$,

where q^{ω} is the Ω-measurable conditional horse lottery given by $q^{\omega}(\omega', x) := q(\omega, x)$ for every $(\omega', x) \in \Omega \times \mathcal{X}$. This axiom has been considered for instance by Galaabaatar and Karni (2013), and it is a weaker version of Savage's postulate P7. We have shown that it is related to independent products of marginal models in the theory of coherent lower previsions, and more particularly to the *strong product* (Zaffalon and Miranda 2017, Section 5.3).

Note, however, that the result does not extend immediately to the case where both Ω, $\mathcal{X}$ are infinite; as shown by Miranda and Zaffalon (2015a, Example 1), independent products (and in particular the strong product) of coherent lower previsions may not even exist in that case. In fact, we find more natural to rely on a weaker, asymmetric, notion of irrelevance that always exists and that suffices to achieve the wanted decomposition of beliefs and values. We illustrate this in "Appendix A" following ideas from our previous work (Zaffalon and Miranda 2017).

3.4 Probability and utility

We have established an equivalence result for desirability and preference at the level of the cone of desirable gambles. This allows us to focus on desirable gambles in the rest of the paper, while aiming to address questions of preference.

The result is very general since it allows us to deal also with problems that cannot be addressed with (sets of) expected utilities, that is, with all the problems that critically depend on a rigorous treatment of zero probabilities or expectations. For all the other problems, it is often convenient to work directly with a probability-utility representation. In our case, this means deducing from a coherent set of desirable gambles the corresponding lower previsions.

More precisely, given a coherent set of desirable gambles $\mathcal{R} \subseteq \mathcal{L}(\Omega \times \mathcal{X})$, we obtain its corresponding lower prevision for all $f \in \mathcal{L}(\Omega \times \mathcal{X})$ using (1):

$$\underline{P}(f) := \sup\{\mu \in \mathbb{R} : f - \mu \in \mathcal{R}\}.$$

$\underline{P}(f)$ is interpreted as the lower expectation of f taken with respect to the probabilities and utilities that are implicit in the definition of $\mathcal{R}$. These can be made explicit very simply by applying $\underline{P}$ to some special indicator functions so as to use marginalisation (see Definition 15); we have that:

$$\underline{P}(B \times \mathcal{X}), \underline{P}(\Omega \times C)$$

are respectively the lower probability of $B \subseteq \Omega$ and the lower utility of $C \subseteq \mathcal{X}$. Note that:

- The lower probability corresponds to a set of finitely additive probabilities on Ω, as it follows from (2).
- Similarly, the lower utility corresponds to a set of finitely additive probabilities too, again using (2), which this time are interpreted as utilities. This means that a utility function in our formalism is mathematically equal to a finitely additive probability on $\mathcal{X}$. A discussion on this point can be found in our past work (Zaffalon and Miranda 2017, Section 4.1).
- Separate probabilities and utilities cannot reproduce $\underline{P}$ unless this is subject to the property of 'state independence'; in the opposite condition of state dependence, the analysis needs to be done using the 'joint' model $\underline{P}$ directly. See "Appendix A" for details.

These considerations can be extended to the conditional case in a straightforward way. For instance, we can use (4) to compute the conditional lower prevision of a gamble $f \in \mathcal{L}(\Omega \times \mathcal{X})$ given some $B \subseteq \Omega$:

$$\underline{P}(f|B \times \mathcal{X}) := \sup\{\mu \in \mathbb{R} : (B \times \mathcal{X})(f - \mu) \in \mathcal{R}\}.$$

This can be given the updating interpretation: $\underline{P}(f|B \times \mathcal{X})$ would then represent our lower expected value of f under the assumption that B occurs and that it is the only thing we get to know about Ω. Note that $\underline{P}(\cdot|B \times \mathcal{X})$ updates both our beliefs and values in the general setting of this paper.

Note also that, mathematically speaking, nothing prevents us from computing $\underline{P}(f|\Omega \times C)$: the lower prevision of f conditional on a subset C of prizes. But we cannot give it the updating interpretation as C simply does not 'occur'. See Sect. 6 for more details.

4 Interpreting the rewards

We have shown that there is an equivalence between coherent preferences and coherent sets of desirable gambles. We exploit this fact in order to give a clear interpretation of the process that allows us to win prizes by using the framework of gambles.

In order to avoid confusion, let us remark that the discussion in this section should better be understood as the description of a 'hypothetical' reward process: most probably we will not want to run this process in practice, but we need the intuition it conveys in order to give meaning to rewards. And we need this meaning to be able to see which gambles are desirable to us and which are not. Stated differently, the previous sections have described the theory, while here we talk about its interpretation; we should bear in mind not to mix up these two levels.

The process by which we are rewarded is based on the idea of having compound lotteries, which underlies the definition of conditional horse lotteries.

The basic framework is constituted by a collection of simple lotteries, one per pair $(\omega, x) \in \Omega \times \mathcal{X}$, which we denote by

$$L := \{\ell_{(\omega,x)} : (\omega, x) \in \Omega \times \mathcal{X}\}.$$

In each of these we can either win prize x or nothing. Before the process starts, we have already a $\frac{1}{2}$ chance of winning prize x, for each $x \in \mathcal{X}$ under each $\omega \in \Omega$.

Then the process goes like this:

1. We are offered a finite number of gambles $f_1, \ldots, f_n \in \mathcal{L}(\Omega \times \mathcal{X})$; we select a subset of them that we accept: without loss of generality, say $f_1, \ldots, f_m$, with $m \leq n$.
2. As a consequence, we are given $\varepsilon \sum_{i=1}^{m} f_i$ in probability currency, where $0 < \varepsilon \ll 1/\sum_{i=1}^{n} \sup|f_i|$ is a constant. Our new probabilities to win the prizes then become
$$p := \frac{1}{2} + \varepsilon \sum_{i=1}^{m} f_i. \tag{8}$$
3. Event ω occurs. We focus on the subset of simple lotteries given by
$$L_\omega := \{\ell_{(\omega,x)} \in L : x \in \mathcal{X}\}. \tag{9}$$

   One of the simple lotteries from L_ω is chosen by an extraneous device that is unknown to us, except for the fact that it is independent of us and of our choices, and that we have no reason to believe unfair. Such a lottery is eventually run so as we can win prize x with probability $\rho(\omega, x)$.

The most important part of this process is that the rewards are given in probability currency, because this makes sure that Axioms D1–D4 are automatically satisfied. Let us show why by following the reasoning outlined by Walley (1991, Section 2.2.4):

- Axioms D1 is a trivial case: assume that among $f_1, \ldots, f_n$ there is a positive gamble $f \in \mathcal{L}^+$. We are certainly willing to accept it as by doing so we increase our chance to win the prizes.
- As for Axiom D4, assume that among $f_1, \ldots, f_n$ there are gambles f and λf, with $\lambda > 0$. Assume for the moment that $\lambda \geq 1$. Compare (a) the reward $\varepsilon\lambda f$ in probability currency with (b) a reward $\varepsilon\lambda f$ in probability currency if an extraneous random event C with known positive chance α occurs (C has to be unrelated to the experiments under considerations and more in general to us). Since (b) yields zero if C does not occur, then it should be desirable to us if and only if we desire it conditional on C. In this case (a) and (b) are equivalent, thus we should desire (a) if and only if we desire (b). But (b) is equivalent to a payment in probability currency of $\alpha\varepsilon\lambda f$. This shows that $\varepsilon\lambda f$ is desirable if and only if $\alpha\varepsilon\lambda f$ is desirable for all $\alpha \in (0, 1]$. Choosing $\alpha := 1/\lambda$, we obtain that λf should be desirable if and only if f is desirable. The case where $\lambda < 1$ is analogous.
- With respect to Axiom D2, we first notice that, reasoning in an analogous way to the case of D1, we are certainly not willing to accept a gamble $f \lneq 0$. In order to enforce this constraint, we require that the zero gamble not be acceptable: in this way, in case we accepted the negative gamble f, then $-f$ would be positive and hence accepted through D1, but their sum, accepted via D4, would be zero. By making zero not acceptable, we thus reject the negative gambles. Therefore Axiom D2 follows.[6]
- With respect to D3, assume that among $f_1, \ldots, f_n$ there are gambles f, g that are both desirable. Consider again an extraneous event C with probability $\frac{1}{2}$ and the compound gamble h defined as follows: if C occurs we are given f and otherwise we are given g. Given that h is desirable both if C occurs and if it does not, then h should be unconditionally desirable. But the probability currency corresponding to h is equal to $\frac{1}{2}\varepsilon f + \frac{1}{2}\varepsilon g$, whence $\frac{1}{2} f + \frac{1}{2} g$ should be desirable; applying D4 we obtain that $f + g$ is desirable.

The interpretation above is based on the actual construction of a linear utility scale through the idea of probability currency and using gambles. Using such a scale allows us to have Axioms D1–D4 automatically satisfied. As a side note, observe that all this is granted also by the requirement to use a small ε in the definition of ρ that keeps it far away from the deterministic probabilities: in fact, if we allowed ρ to get to 0 or 1, then knowing that we could win or lose a prize (almost) for sure would invalidate the linear character of our wealth as a function of gamble values—given that we have non-linear utility in the prizes in general.

[6] This is not the only way to reject the negative gambles: e.g., we could accept the zero gamble and directly require in an axiom that the negative gambles not be acceptable. However, we want to deal with strict preferences in this paper and for that we need irreflexivity, that is, D2.

The same ideas and outcomes apply to preferences, given the equivalence between the formalisms of preference and desirability. This means that such a type of rewards satisfy in particular Axioms A3 and A4. What may be worth remarking, in the case of preferences, is that conditional horse lotteries return probabilities that are only proportional to the ones used in the simple lotteries that enable us to win prizes. This is a sensible point to properly interpret horse lotteries, because one could be misled to think that those are the absolute probabilities by which prizes are won. But this view would conflict in particular with the mixture independence axiom A4, that is, with the underlying requirement of linearity. Stated differently, the long-term controversy that surrounds Axiom A4 in the literature (see Allais 1953; Machina 1982), which has eventually given rise to a number of alternative formalisms that drop A4, appears to be originated by using it jointly with an interpretation of horse lotteries that is, indeed, inconsistent with it. This is not to diminish the value of the criticisms of A4 in any way, it is only to remark the importance of having a correct interpretation of a mathematical theory.

Finally, we can also sketch another interpretation of the rewards that is favoured by the use of sets of desirable gambles and that may be somewhat more realistic than that based on probability currency. The idea is that we could regard reward $f(\omega, x)$ as an amount of prize x; it can also be negative, in which case it is a loss of $-f(\omega, x)$ units of x. Similarly to the case of the probability currency, at some point there would be an extraneous device deciding which $x \in \mathcal{X}$ would be selected and we would be rewarded with $f(\omega, x)$ units of x. This kind of interpretation is well posed if and only if, for all $x \in \mathcal{X}$, our wealth is a linear function of the amounts of prize x. This is in fact the rationale behind the desirability axioms D3 and D4.

As a side note, consider that the linear character of the way we desire each prize does not prevent us from desiring some prizes more than others, in general; this is the reason why we need to distinguish the prizes and the related amounts we might get. As an example, we might imagine the prizes to be different currencies (e.g., dollars, euros, francs, etc.); if we bound the amount of money we can win or lose to some small quantity, then it is known that our appreciation of the prizes is approximately linear. At the same time we will obviously desire strong currencies more than weak ones.

5 Horse lotteries versus conditional horse lotteries

The proposed formulation of rational preferences in this paper, as detailed in Sect. 2.2, is slightly different from the traditional one, because the main modelling unit is that of conditional horse lotteries rather than horse lotteries. This leads, in turn, to a slightly different formulation of the rationality axioms. The aim of this section is to clarify the relation between the two formulations, by distinguishing in particular the case of finite vs infinite space of prizes.

5.1 Finite case

We show that when we restrict the attention to the case of a finite set of prizes $\mathcal{X}$, the two mentioned formalisms for preference are equivalent. Let us briefly recall them:

F_1. The first is the traditional development of decision theory, à la Anscombe-Aumann.
In this formulation, the set of prizes is defined by $\mathcal{X}_z := \mathcal{X} \cup \{z\}$, where $z \notin \mathcal{X}$ is meant to represent the null prize: that is, not receiving any prize from $\mathcal{X}$.
The basic object of the formalism is that of a *horse lottery*: it is a function $p_1 : \Omega \times \mathcal{X}_z \to [0, 1]$, with the additional requirement that $\sum_{x\in\mathcal{X}_z} p_1(\omega, x) = 1$ for all $\omega \in \Omega$. In other words, a horse lottery is a collection of probability mass functions $p_1(\omega, \cdot)$, one per each element of the possibility space Ω. The set of all horse lotteries is denoted by $\mathcal{H}_z$.
As usual, one defines a preference relation $\succ$, in this case between horse lotteries. This relation is assumed to have the worst act, which, without loss of generality (from Zaffalon and Miranda 2017, Section 3.1), can be represented by the act z that is identically degenerate on outcome z: $z(\omega, z) := 1$ for all $\omega \in \Omega$. The relation is subject to coherence axioms analogous to A1–A4 so that if we let ψ denote the operator that drops all the z-elements from acts, or differences of them, we get that $\mathcal{R}_1 := \{\lambda\psi(p_1 - q_1) : \lambda > 0, p_1 \succ q_1\}$ is a coherent set of desirable gambles in $\mathcal{L}(\Omega \times \mathcal{X})$ (Zaffalon and Miranda 2017, Section 3.2).
The interpretation of a horse lottery is that of a pair of nested lotteries. At the outer level there is the experiment that determines the element $\omega \in \Omega$ that occurs. This part is in common with conditional horse lotteries. The inner level is different: in the case of a horse lottery, the mass function $p_1(\omega, \cdot)$ is related to a *single* simple lottery. In order to have a consistent interpretation of horse lotteries with that illustrated in Sect. 4 for conditional horse lotteries, we are going to assume what follows:

- Initially we have a uniform probability over winning or losing, i.e., $\frac{1}{2}$ probability to win and $\frac{1}{2}$ to lose, that is, to receive z (note that these are the same initial probabilities we have with conditional horse lotteries). In turn, we assume that the probability to win prizes is uniform as well, and so equal to $k/2 := 1/(2|\mathcal{X}|)$.
- Such a probability changes after receiving $p_1(\omega, \cdot)$, becoming equal to

$$k\left(\frac{1}{2} + \varepsilon p_1(\omega, x)\right) \tag{10}$$

for all $x \in \mathcal{X}$, and with the constant $\varepsilon > 0$ determined so as to make the resulting numbers probabilities (and far away from the deterministic cases, similarly to the discussion in Sect. 4). The probability of z is determined by those of the prizes.

F_2. The second formalism is precisely that defined in Section 2.2, with the given semantics.

- According to Sect. 4, we assume that only one lottery in L_ω, from (9), is selected with probability $k = 1/|\mathcal{X}|$, and then run; and the chance that we win prize x through a certain conditional horse lottery p_2 is equal to

$$k\left(\frac{1}{2}+\varepsilon p_2(\omega, x)\right) \tag{11}$$

for all $x \in \mathcal{X}$, as it follows from (8). (Here we are implicitly assuming, without loss of generality, that p_2 is one of the gambles $f_1, \ldots, f_m$ in Sect. 4.)

Note that we are using the same symbol ε for the constants in (10) and (11). The reason is that even if those two constants were different, we could always take their minimum and use it in both equations. In fact, the absolute value of the constant is not important and we could actually work out the rest of the section with two different constants, but the development would be unnecessary complicated. For this reason, in the rest of the section we take the two constants to be equal.

In the following we use subscripts 1 and 2 for acts in order to denote the specific formalism (F_1 and F_2) an act belongs to. We keep the same symbol $\succ$, though, for the preference relations defined within the different formalisms.

It turns out, perhaps not surprisingly, that formalisms F_1 and F_2 are fully equivalent ways to deal with preferences in the finite case:

Proposition 4 *Let $\mathcal{R}_1$ and $\mathcal{R}_2$ be the two coherent sets of desirable gambles our preferences originate respectively within F_1 and F_2. Then it follows that $\mathcal{R}_1 = \mathcal{R}_2$.*

Yet, we prefer the less traditional formalism F_2 over F_1: for one thing, the axioms it is bound to seem more direct and simple than those of F_1, and the corresponding proofs seem so as well; most importantly, the infinite case appears to be much more natural, if not just possible, to formulate in full generality using conditional horse lotteries.

5.2 Infinite case

We move on now to consider the case where $\mathcal{X}$ can be infinite, while imposing no restrictions on its cardinality.

5.2.1 Formalism F_2

We start by considering formalism F_2, that is, the one used in this paper as detailed in Sect. 2.2.

The essential difference that arises in F_2 compared to the case of finite $\mathcal{X}$ is concerned with the question of selecting one of the lotteries in L_ω, given that this set can be infinite and even unbounded; in other words, the problem is how we should implement step 3 of the reward process (Sect. 4) in the case of infinite $\mathcal{X}$. Section 4 stressed that the selection is supposed to be "fair", because otherwise we would be induced to bias our beliefs and values in order to account for the lottery-sampling process. In the finite case, the fair selection was implemented by sampling lotteries uniformly at random. How about the infinite case?

In the infinite case, uniform selection is much more of a delicate matter. The central issue is that uniform selection cannot be done in general using σ-additive probabilities; finitely additive probabilities can be used to that end, but they are not constructible in general (see, e.g., Walley 1991, Sections 2.9.5–2.9.7 and 3.5.7–3.5.9). One could wonder, in addition, what it means in practice to sample from a finitely additive probability. Stated differently, it is clear that there is no real way in which one can actually be rewarded through (conditional) horse lotteries in the general case of infinitely many prizes in $\mathcal{X}$.

However, what really matters is the viewpoint of the subject that assesses the desirability of the gambles; for this subject there is no need to know the details of the sampling process or that it can actually be implemented; it is enough that the process be regarded as fair to enable an assessment of desirability (and if the adjective 'fair' entails some degree of ambiguity for the subject, then this will eventually be reflected in some additional imprecision in the subject's set of desirable gambles, which will anyway be naturally accommodated by the theory).

In other words, the present discussion serves the purpose to delineate an ideal reward process, which however allows us to have a clear interpretation behind the formalism of preferences/desirability.

To make all this clearer, we briefly rewrite the selection process that allows us to win prizes in more formal terms. We denote by $P_0 : \mathcal{L}(\mathcal{X}) \to \mathbb{R}$ the linear prevision corresponding to a finitely additive probability that selects a lottery in L_ω uniformly at random. Consider the event {win} that a prize is won and its complementary event {lose}, and let $\mathcal{W} := \{\text{win}, \text{lose}\}$. We denote by $P_0(\cdot|\mathcal{X}) : \mathcal{L}(\mathcal{W} \times \mathcal{X}) \to \mathbb{R}$ the separately coherent conditional linear prevision that represents, for all $x \in \mathcal{X}$, our probability of winning prize x once lottery $\ell_{(\omega,x)}$ has been selected; in practice, $P_0(\{\text{win}\}|\{x\})$ is defined by (8). The joint model for the selection of lotteries and prizes is obtained by marginal extension:[7]

$$P_0' := P_0(P_0(\cdot|\mathcal{X})).$$

P_0' is a linear prevision defined on $\mathcal{L}(\mathcal{W} \times \mathcal{X})$ that corresponds to a finitely additive probability (given that so does P_0), and that represents the mechanism by which we are eventually rewarded.

5.2.2 Formalism F_1

Formalism F_1 makes things more complicated when it comes to define the mechanism to assign prizes. The reason is that horse lotteries $p \in \mathcal{H}_z$ are traditionally defined in such a way that $p(\omega, \cdot)$ is a probability over $\mathcal{X}_z$ for all $\omega \in \Omega$. This can be made to work when $\mathcal{X}_z$ is countable by means of countable additive probabilities (see, e.g., the work by Seidenfeld et al. 1995), even though there are limits to it: for instance we cannot represent the situation where for a certain ω we are rewarded with an equal amount of probability currency for all prizes—given that uniform distributions on a countable space cannot be countably additive. The situation becomes even more complex when $\mathcal{X}$ is continuous.

The straightforward idea that comes to mind is then to let $p(\omega, \cdot)$ be a finitely additive probability; this would also cover the case of continuous and unbounded $\mathcal{X}_z$. But

[7] Strictly speaking, P_0 should be defined on the subset of $\mathcal{X}$-measurable gambles of $\mathcal{L}(\mathcal{W} \times \mathcal{X})$. We neglect this detail to make things simpler.

this typically implies $p(\omega, x) = 0$ for all $x \in \mathcal{X}_z$, which makes p hard to specify in a way that is clear at all for the subject that is assumed to take gambles; moreover, finitely additive probabilities are often non-constructible, so even in principle we would be prevented from clearly seeing how a certain act is made. In addition, how would gambles look like? For in the end we need to consider scaled differences of horse lotteries, such as $f := \lambda(\psi(p) - \psi(q))$; but this would be a finitely additive set-function with all the problems mentioned above (e.g., it would be identically equal to zero on the singletons). Note that restricting the attention to a bounded space $\mathcal{X}$ would partially alleviate the problem, since if we were not to use finitely additive probabilities (one might want to use density functions for instance) we would start having measurability problems; so not all gambles (or not all spaces) would be admissible for use. We prefer a theory to be free from this type of constraints in general.

So it seems to us that traditional horse lotteries are not really suited to be extended to handle infinitely many prizes in general. We find it more natural and powerful to move to conditional horse lotteries, which solve the problem right away and in full generality, and, as a bonus, make axioms as well as proofs simpler.

As a final note, consider that all the problems of F_1 are originated by the requirement that $p(\omega, \cdot)$ be a probability. But why should it be? The requirement appears to follow from the traditional interpretation that regards $p(\omega, \cdot)$ as the probabilities by which we would win prizes running the related simple lottery. But we have already argued in Sect. 4 that this interpretation is unfortunate, as it is inconsistent with the linearity assumption embodied by A4 in particular. A more sensible interpretation regards $p(\omega, \cdot)$ only as a relative increase of a pre-existing probability over prizes; as such, it need not be normalised, i.e., be a probability. Once we take this step, it is immediate to consider the opportunity to work with conditional horse lotteries; so a correct interpretation favours also a more natural formulation.

6 Conglomerability

Let us start by considering an uncertainty model alone, for the moment neglecting considerations of value. We recall the formulation of conglomerability given in Definition 12:

$$(\forall f \in \mathcal{L}(\Omega))((\forall B \in \mathcal{B})\ Bf \in \mathcal{R} \cup \{0\} \Rightarrow f \in \mathcal{R} \cup \{0\}).$$

Conglomerability follows from additivity (D3) in case $\mathcal{B}$ is finite; therefore it is an issue only for infinite partitions. The debate about whether or not conglomerability should be imposed in that case has not been settled in spite of the long time passed since de Finetti's (1931) seminal work. The controversy seems to be kept alive by the tension originating from two opposite tendencies: on the one hand, the fact that imposing conglomerability makes things mathematically quite harder (the corresponding models being somewhat less well behaved); on the other, that the absence of conglomerability creates a number of paradoxical situations that seem to require some fixing.

In the following we address the question from a specific point of view, that of 'temporal coherence', thus without pretending to be exhaustive on the matter. For a broader discussion about conglomerability, see, e.g., Berti et al. (2017). For some logical and philosophical perspectives, see Howson (2009) and DiBella (2018).

6.1 The scope of conglomerability in the case of beliefs

In some past work we have established a clear connection between conglomerability and the question of being coherent in time (Zaffalon and Miranda 2013).

Assume that we hold beliefs at present time in the form of a coherent set of desirable gambles $\mathcal{R}$. At some later time an event $B \in \mathcal{B}$ occurs; we will then hold new beliefs in the form of a coherent set of desirable gambles $\mathcal{R}^B \subseteq \mathcal{L}(B)$. $\mathcal{R}$ and $\mathcal{R}^B \subseteq \mathcal{L}(B)$ need not entertain any relation; in particular, $\mathcal{R}^B$ need not be the set of conditional beliefs that can be derived from $\mathcal{R}$ and that are denoted by $\mathcal{R}\rfloor B \subseteq \mathcal{L}(B)$ (this point appears to have been explicitly raised first by Hacking 1967). To clarify this point, let us recall the following:

Definition 23 (*Perfect information*) If $B \in \mathcal{B}$ represents all and only the information we receive about Ω from the establishment of our current beliefs $\mathcal{R}$ up to the occurrence of B, then we say we are in a case of *perfect information* (this terminology has been introduced by Shafer et al. 2003).

Thanks to this definition, we can precisely recall what updated beliefs mean by the next definition:

Definition 24 (*Updating interpretation—beliefs*) Conditional beliefs $\mathcal{R}\rfloor B$ are called *updated beliefs* when they are interpreted as beliefs under the assumption that B occurs in a case of perfect information.

Note in particular that updated beliefs are still beliefs expressed at present time, so in principle they have no relation with future beliefs, those that we hold *after* the occurrence of B.

However, there are situations where $\mathcal{R}^B$ and $\mathcal{R}\rfloor B$ are related. One of these cases arises when we decide to establish our future 'beliefs' at present time (in this case, we we will talk of future *commitments* rather than beliefs). This means that we decide in advance to stick to the future behaviour that is induced by current conditional beliefs; in other words, we decide in advance that $\mathcal{R}^B := \mathcal{R}\rfloor B$ for all $B \in \mathcal{B}$.

When may this be going to be the case?

C1. A condition that seems necessary is that we are in a case of perfect information, otherwise conditional beliefs would not represent our actual beliefs after B occurs as these would involve considerations about evidence other than B.

C2. What appears to be left is our capability to examine the evidence at hand to the extent of establishing our current beliefs as well as we can. If we have enough time and any other needed resource to that end, we can assume that our uncertainty model is the best, in our judgement, we can produce on the basis of the given evidence.

Given these two conditions, there seems to be no reason why we might want a model for future beliefs different from current conditional ones.

But establishing future commitments at the time of present beliefs gives us the opportunity to check, and actually impose as a rationality requirement, that the two of them cohere with each other. Note that this is an additional property, required on top of the (separate) coherence of $\mathcal{R}$ and the coherence of $\mathcal{R}^B$ (for all $B \in \mathcal{B}$), that arises out of temporal considerations. We have called it *strong temporal coherence*.[8]

The main outcome of our 2013 work, in this regard, is that strong temporal coherence coincides with the conglomerability of $\mathcal{R}$.

For us, this result essentially closes the long controversy about conglomerability: conglomerability turns out to be a form of temporal coherence, which means that it is a rationality requirement when we decide in advance to establish future commitments equal to current conditional beliefs. In case we are not interested in future beliefs, for instance because we work with unconditional models only, then conglomerability is not a rationality requirement for an uncertainty model.

However, the scope of conglomerability is still quite narrow. This happens in particular because of condition C2, as it requires the availability of resources we may as well not have, first of all the time to examine all the evidence carefully. What we are going to argue now is that there is a simple, realistic, assumption that can render the scope of conglomerability much wider, by essentially removing the need of C2. It is based on the following:

Definition 25 (*Reliable assessments*) We say that an assessment is *reliable* if it is not stronger than what is justified by the evidence and the available resources, including time. Similarly, we say that the assessment process is reliable if all assessments are.

The concept of reliable assessments appears to have been introduced first by Walley (1991, Sect. 6.1.2).

It is important to realise, first of all, that the assessment process can always be reliable with imprecise probability models: the reason is that in such a setting we can weaken our assessments as much as we need, up to expressing a condition of total lack of knowledge. So what Definition 25 describes is simply a process that does not yield overconfident assessments; this can always be achieved by acting on their strength. We can even weaken our assessments some extra bit in case we prefer to be very confident not to express too strong judgements (in this case, we would yield an *incomplete representation* of our beliefs, rather than an *exhaustive* one; see Walley 1991, Section 2.10.3 for a discussion about this point).

In any case the point is that the assumption of reliability implies that we are not taking appreciable risks in the assessment process while modelling the evidence as well as we can, given the available resources. For this reason, we should definitely be willing to set future commitments equal to current conditional beliefs. But once we do so, we can argue again, as we have done in our 2013 work, that conglomerability is a consequence of strong temporal coherence. To summarise the situation, what we argue is the following:

[8] For a discussion of the rationality of this axiom and the connection with other, weaker, requirements, see (Zaffalon and Miranda 2013, Section 6.4).

Whenever we are interested in future behaviour in a setting of perfect information, and our assessments are reliable, then conglomerability is a rationality requirement.

And since the assumption of reliability can always be met (it is just a 'good practice' in the assessment process), it makes the scope of conglomerability very wide: it is essentially extended to all cases of perfect information—which remains then the only actual constraint.

An important remark is that all this does not prevent us from behaving in the future according to a probabilistic model $\mathcal{R}'^B$ that is more precise than that characterised by our conditional beliefs, i.e., such that $\mathcal{R}\rfloor B = \mathcal{R}^B \subseteq \mathcal{R}'^B$; there is not any form of incoherence in doing so. In fact the meaning of a probabilistic model is to prescribe the limits of what we can do, but within those limits we have ample choice; we can in fact behave as we prefer as long as the model that we use to that end contains $\mathcal{R}^B$ (this constraint has to be maintained since it represents our established future commitments). This is useful and important given that there might be a certain time between the establishment of conditional beliefs and the future occurrence of B. This may make us reflect more deeply about the evidence, thanks possibly also to the availability of resources other than time (such as computational resources), and make us see that we can be somewhat bolder in our evaluations.

Finally, let us recall that so far we have been considering the case of imprecise probability. If we instead restrict the attention to the more traditional case of precise, Bayesian, probability, it turns out that the assumption of reliability is not needed: conglomerability is a rationality requirement in a temporal setting based on perfect information alone (Zaffalon and Miranda 2013, Section 6.5). The rationale is that, for a Bayesian violating reliability amounts to violating the idea of being Bayesian itself, since in that case Bayes' rule is the only temporal-coherent way to update beliefs; whence sticking to reliability can be assumed outright for a Bayesian.

6.2 Extension to the general case of beliefs and values

Now we turn to the general case where $\mathcal{R} \subseteq \mathcal{L}(\Omega \times \mathcal{X})$. It is important to remember, from Sect. 3.4 in particular, that Ω is still the only space of possibilities: the elements of $\mathcal{X}$ do not 'occur'; as a consequence it only makes sense to update $\mathcal{R}$ on subsets of Ω alone.

Having made this point, now we proceed with our focus on updating, and in particular on the definition of conglomerability extended so as to jointly handling the case of beliefs and values.

Definition 26 (*Conglomerability—extended*) Consider a coherent set of desirable gambles $\mathcal{R} \subseteq \mathcal{L}(\Omega \times \mathcal{X})$ and let $\mathcal{B}$ be a partition of Ω. $\mathcal{R}$ is said to be *$\mathcal{B}$-conglomerable* if it satisfies the following condition:

D5. $(\forall f \in \mathcal{L}(\Omega \times \mathcal{X}))((\forall B \in \mathcal{B})\ (B \times \mathcal{X})f \in \mathcal{R} \cup \{0\} \Rightarrow f \in \mathcal{R} \cup \{0\})$ [Conglomerability].

The rationale behind this definition is exactly as before in Definition 12. Moreover, the mathematical form is also as before: we are simply focusing on a special type of partitions in the enlarged space $\Omega \times \mathcal{X}$. For this reason, the conclusions of our 2013 work apply here as well, as summarised in the previous section.

The additional discussion concerned with reliability and perfect information essentially applies too; there is however a caveat in this general case, which has to do with the different nature of beliefs and values. Once we make our joint assessments of beliefs and values in a reliable way (Definition 25), we still create the conditions, as before, to stick to the future commitments $\mathcal{R}^{B\times\mathcal{X}}$ in the form of our current conditional assessments $\mathcal{R}\rfloor(B \times \mathcal{X})$. But the time that passes from the establishment of $\mathcal{R}$ to the occurrence of B may have a different effect on our values compared to our beliefs. In fact, under perfect information, it is perfectly plausible that our beliefs can only become stronger in time. This is not the case of our values, which might change in unpredictable ways due to the occurrence of events that have nothing to do with B or $\Omega \times \mathcal{X}$: we could for instance become very risk-averse as a consequence of a financial loss. For this reason, it seems necessary to require in addition what follows:

Definition 27 (*Perfect isolation*) If $B \in \mathcal{B}$ represents the only information that can affect our values in the experiment concerned with $\Omega \times \mathcal{X}$, then we say we are in a case of *perfect isolation* of values from external factors.

Note that, as a consequence, the updating interpretation should be extended to account for isolation in the general case of beliefs and values:

Definition 28 (*Updating interpretation*) Conditional assessments $\mathcal{R}\rfloor(B \times \mathcal{X})$ are called *updated assessments* when they are interpreted as assessments under the assumption that B occurs in a case of perfect information and isolation.

Having said this, the situation can finally be summarised as follows:

> In the general case, conglomerability is a consequence of reliability, perfect information and isolation.

Under these assumptions (and of course in a temporal setting) D5 should be added to D1–D4 as a further rationality requirement on preferences.

Note, as discussed in the previous section, that reliability can be skipped when we are Bayesian (in this case: precise beliefs and values); whence conglomerability is a consequence of perfect information and isolation only.

6.3 A final remark

We have argued that conglomerability should be a rationality requirement under very broad conditions. This follows from the identification of conglomerability with a special type of coherence: if conglomerability is not imposed, we can find situations where we are inconsistent with ourselves at the very same point in time—in spite of the fact that some of our beliefs and values are regarded as commitments for the future. This is clearly a principled support for the adoption of conglomerability.

There is, however, also a very practical support for conglomerability that comes along with the principled justification; it is the fact that conglomerability can make our inferences very strong, and hence informative, compared to the case where we do not impose it. Consider for instance Example 4 in Miranda and Zaffalon's (2015a) work. There we show that a model that is vacuous between two possibilities, that is, completely uninformative about them, becomes a precise probabilistic model after it is corrected to make it conglomerable. In this case the sole adoption of conglomerability makes such a change possible. We see that conglomerability can have a very practical role to play in our models that goes well beyond its theoretical justification.

7 Conclusions

In this paper we have reconsidered our previous work on the equivalence between (incomplete) rational preferences and desirability.

We have shown that such an equivalence continues to hold when we formulate rational preferences in full generality, with unconstrained spaces of possibilities and prizes, unlike in our original work where prizes were in a finite number. Moreover, we have discussed the role of conglomerability and deduced that it should be a rationality requirement under weak assumptions, and whence it should be widely imposed in probabilistic and decision-theoretic models.

We have also provided a rigorous interpretation of gamble values based on the idea of probability currency. This allows us to make desirability live on its own, without the need to base its development on preferences, in the sense that it becomes a self-contained theory. As a consequence, one may wonder whether the founding notion for decision-theoretic modelling should be that of desirability rather than preference. In fact we are inclined to regard desirability as the primary notion, given that both the axioms and the theory itself appear to become much more direct and easy to access.

All this provides the foundations for a very general theory of uncertainty and decision making. Where can we go from here?

There are many directions that we envisage for future research. A very natural extension of the ideas presented here involves considering the recent work by Van Camp et al. (2018a) (see also Van Camp 2018). It is based on so-called *coherent choice functions*, which generalise preference modelling to non-binary comparisons of options (the original axiomatisation for the case of incomparability is due to Seidenfeld et al. 2010, with slightly different axioms). This gives additional modelling power compared to incomplete preferences and desirability, while including these models as particular cases (Van Camp et al. 2018a, Section 4). Notably, Van Camp et al. (2018a, Section 3) have already exploited our 2017 work so as to extend choice functions to handle values besides beliefs, albeit in the case of finitely many prizes. The present paper could allow such an approach to be eventually generalised to the case of fully unconstrained spaces.

A different opportunity of research lies instead in the connection established by Benavoli et al. (2016a) between desirability and quantum mechanics: in short, the authors prove that imposing D1–D4 on gambles made of Hermitian matrices is enough to derive quantum mechanics, or, stated differently, that quantum mechanics is nothing else than desirability in an opportune domain. Subsequent work by Benavoli et al. (2016b) has extended those ideas relying on our 2017 work so as to derive a rational theory of decision making for quantum experiments. All this could be, again, combined with the present paper to get to a higher level of generality.

A more pragmatic, though equally important, avenue of research would be to establish clear connections between the theory presented here and all the research in AI based on preferences. It seems that such developments in AI and the theory of desirability have proceeded independently so far; linking them would provide benefits in both directions and would automatically provide a formal basis for robust methods in AI.

Most of all, however, we believe that the traditional theories of preference fall short in the realism of the problems that they are able to address. On the one hand, Anscombe and Aumann's theory is well founded if we adopt the correct interpretation, but it requires paying rewards indirectly, through the idea of probability currency, which entails a certain degree of artificiality; the same happens for the theories that build on it, like the theory we present here. Savage (1954) models rewards directly, but his theory suffers from an underlying assumption of linearity, much in the spirit of Axiom A4, that clashes with the non-linearity of prize values and that eventually leads to Allais' paradox. A number of attempts have been made to correct these issues in the direction of more realistic theories (e.g., see Machina 1982; Tversky and Kahneman 1992; Nau 2011; Cerreira-Vioglio et al. 2016), but the situation still appears quite open. The challenge, in our view, would be to work out a desirability-based theory that can bypass those problems. We believe this should be possible, and we leave it for future research.

Acknowledgements The authors are grateful to the anonymous referees for a careful reading of the paper and in particular to Referee 1 for having spotted some technical problems that are now fixed.

A Decomposition and completeness of preferences

We now turn to the formalisms of desirability and coherent lower previsions to briefly show how they can represent and generalise well-known concepts in rational preferences. The following discussion is a summary of the one we presented in Zaffalon and Miranda (2017, Section 5), with some additional remarks and proofs for the case of infinitely many values.

A.1 In terms of sets of desirable gambles

We begin by considering the case where our assessments are modelled by sets of desirable gambles.

A.1.1 State independence

State independence is the condition that allows us to have separate models for beliefs and values.

We showed in Zaffalon and Miranda (2017, Section 5.3) that the traditional notion of state independence in the literature of preferences is equivalent to the notion of 'strong independence' in imprecise probability. In the case of totally ordered Archimedean preferences (i.e., precise probability and utility models), this means that probability and utility are stochastically independent models, when utility is mathematically represented through a probability function (like we do in this paper, see Sect. 3.4).

Such a notion of independence extends over multiple probability-utility pairs, and even to sets of desirable gambles. However, it ceases to exist when both Ω and $\mathcal{X}$ are infinite (Miranda and Zaffalon 2015a, Example 1).[9] This holds also for most other notions of independence while, interestingly, that need not be the case for notions of 'irrelevance': these are asymmetric notions where mutual (symmetric) independence is replaced by independence in only one direction. In the case of this paper, for instance, this means assuming that values are independent of states but not vice versa; we say that states are irrelevant to values.

This makes of irrelevance a very natural candidate for a notion of state independence that exists no matter the cardinalities of the spaces involved. But there is more to it, as it is arguable that irrelevance qualifies itself as the right notion to use in the present context, unlike independence. The reason is that while it makes sense that states may be irrelevant to values, the opposite seems to be questionable: in fact, as we have already remarked, it appears meaningless to update a model on elements of $\mathcal{X}$, which do not occur; as a consequence the irrelevance of values to states appears meaningless too.

Let us then consider the following definition of irrelevance for desirability:

Definition 29 (*Ω-$\mathcal{X}$ irrelevant product for gambles*) A coherent set of gambles $\mathcal{R}$ on $\Omega \times \mathcal{X}$ is called an *Ω-$\mathcal{X}$ irrelevant product* of its marginal sets of gambles $\mathcal{R}'_\Omega, \mathcal{R}'_\mathcal{X}$ if it includes the set

$$\mathcal{R}|\Omega := \{h \in \mathcal{L}(\Omega \times \mathcal{X}) : (\forall \omega \in \Omega)\ h(\omega, \cdot) \in \mathcal{R}'_\mathcal{X} \cup \{0\}\} \setminus \{0\}.$$

We have already considered these products, in another context and for lower previsions, in Miranda and Zaffalon (2015a) (see also the work by de Cooman and Miranda 2009; de Bock and de Cooman 2014, 2015b).

The rationale of the definition is the following. First, the requirement that $h(\omega, \cdot) \in \mathcal{R}'_\mathcal{X} \cup \{0\}$ is there so that the inferences conditional on $\{\omega\}$ encompassed by $\mathcal{R}|\Omega$ should yield the marginal set $\mathcal{R}'_\mathcal{X}$. This is just a way to formally state that ω is irrelevant to $\mathcal{X}$. The same is repeated for every ω, so that $\mathcal{R}|\Omega$ can be regarded as being born out of aggregating all the irrelevant conditional sets.[10] Finally, that $\mathcal{R}$ contains $\mathcal{R}|\Omega$ is imposed to make sure that $\mathcal{R}$ is a model coherent with the irrelevance of beliefs on values.

If $\mathcal{R}$ satisfies Definition 29 we say it models *state-independent preferences*. The least-committal among these models is given by[11]

---

[9] This is due to the involvement of conglomerability in the definition of independence; de Bock (2017) has recently shown that some other notions of independence continue to exist by dropping conglomerability.

[10] Given that Ω is possibly infinite, this entails an assumption of conglomerability; without such an assumption one could obtain uninformative preferences, see Miranda and Zaffalon (2015a, Section 5.1) for details.

[11] We refer to the work by Moral (2005, Section 2.4), de Cooman and Miranda (2012, Theorem 13) and de Bock and de Cooman (2015b, Theorem 8) for somewhat related results on finite referential spaces.

Proposition 5 *Given two marginal coherent sets of desirable gambles $\mathcal{R}_\Omega, \mathcal{R}_\mathcal{X}$ defined on the subsets of $\mathcal{L}(\Omega \times \mathcal{X})$ given by the Ω-measurable and $\mathcal{X}$-measurable gambles, respectively, their smallest Ω-$\mathcal{X}$ irrelevant product is given by the marginal extension of $\mathcal{R}_\Omega$ and $\mathcal{R}|\{\omega\} := \mathcal{R}_\mathcal{X}$ for all $\omega \in \Omega$. Namely,*

$$\hat{\mathcal{R}} := \{g + h : g \in \mathcal{R}_\Omega \cup \{0\}, h \in \mathcal{R}|\Omega \cup \{0\}\} \setminus \{0\}. \tag{12}$$

Definition 29 and Eq. (12) represent our main proposal to model state-independent preferences.

A.1.2 State dependence

We say that a set of desirable gambles $\mathcal{R}$ models state-dependent preferences when it does not satisfy the conditions in Definition 29, i.e., when it does not include the set $\mathcal{R}|\Omega$ derived from its marginal $\mathcal{R}_\mathcal{X}$ and the assumption of Ω-$\mathcal{X}$ irrelevance.

A.1.3 Completeness

So far we have discussed the general case of coherent sets of gambles $\mathcal{R}$ without discussing in particular the so-called complete or maximal coherent sets for which it holds that $f \notin \mathcal{R} \Rightarrow -f \in \mathcal{R}$ for all $f \in \mathcal{L}$, $f \neq 0$. Given the equivalence between coherent sets of gambles in $\mathcal{L}(\Omega \times \mathcal{X})$ and coherent preferences in $\mathcal{H} \times \mathcal{H}$, discussing the case of maximal sets amounts to discussing the case of complete preferences.

Definition 30 (*Completeness of beliefs and values for gambles*) A coherent set of desirable gambles $\mathcal{R} \subseteq \mathcal{L}(\Omega \times \mathcal{X})$ is said *to represent complete beliefs* if $\mathcal{R}_\Omega$ is maximal. It is said *to represent complete values* if $\mathcal{R}_\mathcal{X}$ is maximal. Finally, if $\mathcal{R}$ is maximal, then it is said *to represent complete preferences.*

The rationale of this definition is straightforward: when we say, for instance, that $\mathcal{R}$ represents complete beliefs, we mean that there is never indecision between two options that are concerned only with Ω. The situation is analogous in the case of complete values. Finally, in the case of complete preferences, the definition is just the direct application of the definition of maximality.

A.2 In terms of lower previsions

We now consider the case where preferences are modelled by means of a coherent lower prevision $\underline{P}$ on $\mathcal{L}(\Omega \times \mathcal{X})$.

A.2.1 State independent preferences

We proceed as in the case of desirability by focusing on irrelevance of states to values.

Definition 31 (*Ω-$\mathcal{X}$ irrelevant product for lower previsions*) Given marginal coherent lower previsions $\underline{P}'_\Omega, \underline{P}'_\mathcal{X}$, let

$$\underline{P}(f|\{\omega\}) := \underline{P}'_\mathcal{X}(f(\omega,\cdot)),$$

for all $f \in \mathcal{L}(\Omega \times \mathcal{X})$, and

$$\underline{P}(\cdot|\Omega) := \sum_{\omega\in\Omega} I_{\{\omega\}}\underline{P}(\cdot|\{\omega\}).$$

Then a coherent lower prevision $\underline{P}$ on $\mathcal{L}(\Omega \times \mathcal{X})$ is called an *Ω-$\mathcal{X}$ irrelevant product* of $\underline{P}'_\Omega, \underline{P}'_\mathcal{X}$ if

$$\underline{P} \geq \underline{P}'_\Omega(\underline{P}(\cdot|\Omega)).$$

In this case we also say that $\underline{P}$ models state-independent preferences.

Here $\underline{P}(\cdot|\Omega)$ plays the role that $\mathcal{R}|\Omega$ took in Section A.1, that is, conglomerating all the conditional information. The concatenation $\underline{P}'_\Omega(\underline{P}(\cdot|\Omega))$ is a marginal extension: in particular, it is the least-committal coherent Ω-conglomerable model built out of the given marginals and the assessment of irrelevance that defines $\underline{P}(\cdot|\Omega)$ through $\underline{P}'_\mathcal{X}$. Every coherent lower prevision that dominates $\underline{P}'_\Omega(\underline{P}(\cdot|\Omega))$ is compatible with the irrelevance assessment but is also more informative than that:

Proposition 6 *Given two marginal coherent lower previsions $\underline{P}'_\Omega, \underline{P}'_\mathcal{X}$, their smallest Ω-$\mathcal{X}$ irrelevant product is given by the marginal extension*

$$\underline{P}'_\Omega(\underline{P}(\cdot|\Omega)).$$

We refer to the work of Walley (1991, Section 6.7), Miranda and de Cooman (2007), de Cooman and Miranda (2009) and de Cooman and Hermans (2008) for additional information on the marginal extension and its use in a number of different contexts.

A.2.2 State dependent preferences

As in the case of desirable gambles, we define state dependence as the lack of state independence: this means that if we consider a coherent lower prevision $\underline{P}$ with marginals $\underline{P}'_\Omega, \underline{P}'_\mathcal{X}$, it is said to model *state-dependent preferences* when it does not dominate the concatenation $\underline{P}'_\Omega(\underline{P}(\cdot|\Omega))$, where $\underline{P}(\cdot|\Omega)$ is derived from $\underline{P}'_\mathcal{X}$ by an assumption of epistemic irrelevance.

A.2.3 Completeness

The definition of completeness for lower previsions is a rephrasing of that for desirable gambles:

Definition 32 (*Completeness of beliefs and values for lower previsions*) A coherent lower prevision $\underline{P}$ on $\mathcal{L}(\Omega \times \mathcal{X})$ is said *to represent complete beliefs* if its marginal $\underline{P}_\Omega$ is linear. It is said *to represent complete values* if its marginal $\underline{P}_\mathcal{X}$ is linear. Finally, if $\underline{P}$ is linear, then it is said *to represent complete preferences*.

There are a number of equivalent ways in which we can characterise the linearity of previsions in terms of the corresponding set of desirable gambles:

Proposition 7 *Let $\underline{P}$ be a coherent lower prevision on $\mathcal{L}$ and $\mathcal{R}$ its corresponding coherent set of strictly desirable gambles. The following are equivalent:*

(i) $\underline{P}$ is a linear prevision.
(ii) If $f \notin \mathcal{R}$ then $\varepsilon - f \in \mathcal{R}$ for all $\varepsilon > 0$.
(iii) $\mathcal{R}$ is negatively additive, meaning that $f, g \notin \mathcal{R} \Rightarrow (\forall \varepsilon > 0)\ f + g - \varepsilon \notin \mathcal{R}$.

We can then focus on any of those formulations to immediately deduce characterisations for all the cases of completeness in preferences:

Proposition 8 *Consider a coherent preference relation $\succ$ on gambles represented by a coherent lower prevision $\underline{P}$ on $\mathcal{L}(\Omega \times \mathcal{X})$, whose corresponding coherent set of gambles is denoted by $\mathcal{R}$. Let $\mathcal{R}_\Omega, \mathcal{R}_\mathcal{X}$ denote its marginals. Then:*

(i) $\underline{P}$ represents complete beliefs $\Leftrightarrow \mathcal{R}_\Omega$ is negatively additive.
(ii) $\underline{P}$ represents complete values $\Leftrightarrow \mathcal{R}_\mathcal{X}$ is negatively additive.
(iii) $\underline{P}$ represents complete preferences $\Leftrightarrow \mathcal{R}$ is negatively additive.

B Proofs

Proof of Proposition 1 Using the mixture-independence axiom A4 together with the assumption,

$$p \succ q \Leftrightarrow \alpha p + (1-\alpha)r \succ \alpha q + (1-\alpha)r = \alpha p + (1-\alpha)s \Leftrightarrow r \succ s.$$

Proof of Proposition 2 By $p \gneq q$ we deduce that $0 \lneq p - q \leq 1$. Then $(p-q) \in \mathcal{H}$ and $(p-q) \succ 0$ by A1. And since

$$\frac{1}{2}(p-q) = \frac{1}{2}[(p-q) - 0],$$

we deduce by Proposition 1 that $p \succ q$.

Proof of Lemma 1 Consider a set of desirable gambles $\mathcal{D} \subseteq \mathcal{L}_1(\Omega \times \mathcal{X})$ and define the linear relation $\succ \subseteq \mathcal{H} \times \mathcal{H}$ by $p \succ q$ if and only if $p - q \in \mathcal{D}$.

Observe also that for any $f \in \mathcal{D}$ we can always find $p, q \in \mathcal{H}$ such that $f = p - q$: it is sufficient to let $p := \max(f, 0)$ and $q := \min(0, f)$.

Let us now prove injectivity. Assume by contradiction that there are sets $\mathcal{D}_1 \neq \mathcal{D}_2$ in $\mathcal{L}_1$ such that the respective linear relations are equal: $\succ_1 = \succ_2$. Since the sets are different, then we can assume without loss of generality that there is $f_1 \in \mathcal{D}_1 \setminus \mathcal{D}_2$. At the same time for all $p, q \in \mathcal{H}$ with $p-q = f_1$, we have that $p \succ_1 q$ and $p \succ_2 q$. This implies that there is $f_2 \in \mathcal{D}_2$ such that $f_2 = p-q$. But then $f_1 = f_2$, a contradiction.

Let us focus on surjectivity. Consider a linear relation $\succ \subseteq \mathcal{H} \times \mathcal{H}$ and let $\mathcal{D}' := \{f \in \mathcal{L}_1(\Omega \times \mathcal{X}) : (\exists p, q \in \mathcal{H})\ f = p - q, p \succ q\}$. The relation it induces is given by $p \succ' q \Leftrightarrow p - q \in \mathcal{D}'$. Now, if $p \succ q$, then $p - q \in \mathcal{D}'$, whence $p \succ' q$. Conversely, if $p \succ' q$, then $p - q \in \mathcal{D}'$; as a consequence, there are $p', q' \in \mathcal{H}$, $p' \succ q'$ such that $p - q = p' - q'$. But relation $\succ$ is linear, whence $p \succ q$. It follows that $\succ = \succ'$.

Proof of Theorem 1 Consider a set of desirable gambles $\mathcal{D} \subseteq \mathcal{L}_1$ and its corresponding linear relation $\succ$. They are related through

$$\mathcal{D} = \{f \in \mathcal{L}_1(\Omega \times \mathcal{X}) : (\exists p, q \in \mathcal{H})\ f = p - q, p \succ q\},$$
$$p \succ q \Leftrightarrow p - q \in \mathcal{D}.$$

(A1 $\Leftrightarrow$ D1′) If $p \succ 0$ for all $p \in \mathcal{H}$, $p \neq 0$, then $p \in \mathcal{D}$; and since $\mathcal{L}_1^+ = \mathcal{H}\backslash\{0\}$, then we have the thesis. Conversely, if $\mathcal{L}_1^+ \subseteq \mathcal{D}$, then $p \in \mathcal{D}$ for all $p \in \mathcal{H}\backslash\{0\}$; as a consequence $p \succ 0$ for all $p \in \mathcal{H}$, $p \neq 0$.

(A2 $\Leftrightarrow$ D2′) By contradiction, if $p \succ p$ for some $p \in \mathcal{H}$, then $0 \in \mathcal{D}$; and if $0 \in \mathcal{D}$, then for some $p \in \mathcal{D}$, $p \succ p$.

(A3 $\Leftrightarrow$ D3′) Assume that the relation is transitive and take $f, g \in \mathcal{D}$ such that $f = p - q$, $g = q - r$ for some $p, q, r \in \mathcal{H}$. Then we have that $p \succ q$ and $q \succ r$; by transitivity we conclude that $p \succ r$, whence $f + g \in \mathcal{D}$. Conversely, assume that $\mathcal{D}$ is additive and take $p \succ q$, $q \succ r$. Letting $f := p - q$, $g := q - r$, we get by additivity that $f + g = p - r \in \mathcal{D}$, so that $p \succ r$.

(A4 $\Leftrightarrow$ D4′) Assume that $\succ$ satisfies mixture independence and take $f \in \mathcal{D}$ and $\lambda > 0$ such that $\lambda f \in \mathcal{L}_1$. Then there are $p \succ q$ such that $f = p - q$. Moreover, there must be $r, s \in \mathcal{H}$ such that $\lambda f = r - s$ given that $\lambda f \leq 1$ (it is enough as usual to take the positive and negative parts of λf, respectively). Using Proposition 1, we get that $r \succ s$, whence $\lambda f \in \mathcal{D}$.
Conversely, assume that $f \in \mathcal{D}$ implies that $\lambda f \in \mathcal{D}$ for all $\lambda > 0$ such that $\lambda f \in \mathcal{L}_1$. Taking $p \succ q$, we have then that $\lambda p + (1 - \lambda) r \succ \lambda q + (1 - \lambda) r$ for all $p, q, r \in \mathcal{H}$ and $\lambda \in (0, 1]$. On the other hand, if $\lambda p + (1 - \lambda) r \succ \lambda q + (1 - \lambda) r$ for all $p, q, r \in \mathcal{H}$ and $\lambda \in (0, 1]$, then trivially $p \succ q$.

(A0 $\Leftrightarrow$ D0′) Assume that for all $p \succ q$, with $p \gneq q$ not holding, there is $\alpha \in (0, 1)$ such that $\alpha p \succ q$. We want to show that for all $f \in \mathcal{D}\backslash\mathcal{L}_1^+$ and all $g \in \mathcal{L}_1$ such that $\max(0, -f) \leq g \leq \min(1, 1 - f)$, there is $\alpha \in (0, 1)$ such that $\alpha f - (1 - \alpha) g \in \mathcal{D}$.
Let $q := g$ and $p := f + q$, so that $f = p - q$. Let us show that $q \in \mathcal{H}$; it is enough to prove that $\max(0, -f) \leq \min(1, 1 - f)$. In case $f(\omega, x) \geq 0$, the inequality reduces itself to $0 \leq 1 - f(\omega, x)$, which is true since $f(\omega, x) \leq 1$. In case $f(\omega, x) < 0$, the inequality becomes $-f(\omega, x) \leq 1$, which is again true since $f(\omega, x) \geq -1$.

Let us similarly show that $p = f + q \in \mathcal{H}$ too. We know that $f + \max(0, -f) \leq p \leq f + \min(1, 1 - f)$. Again, if $f(\omega, x) \geq 0$, we see that $0 \leq f(\omega, x) \leq p(\omega, x) \leq 1$; if $f(\omega, x) < 0$, we have that $0 \leq p(\omega, x) \leq f + 1 \leq 1$.
At this point we can apply the hypothesis, recalling that $f \notin \mathcal{L}_1^+$ and hence that $p \gneq q$ does not hold, and so we deduce that there is $\alpha \in (0, 1)$ such that $\alpha p - q = \alpha(p - q) - (1 - \alpha)q = \alpha f - (1 - \alpha)q = \alpha f - (1 - \alpha)g \in \mathcal{D}$.
Conversely, suppose that for all $f \in \mathcal{D}\backslash\mathcal{L}_1^+$ and all $g \in \mathcal{L}_1$ such that $\max(0, -f) \leq g \leq \min(1, 1 - f)$, there is $\alpha \in (0, 1)$ such that $\alpha f - (1 - \alpha)g \in \mathcal{D}$. Let us show that $p \succ q$, with $p \gneq q$ not holding, implies that there is $\alpha \in (0, 1)$ such that $\alpha p \succ q$.
Let $f := p - q$, from which it follows that $f \notin \mathcal{L}_1^+$, and $g := q$. Observe that if $\max(0, -f) \leq g \leq \min(1, 1 - f)$ holds, then from the assumption the thesis follows immediately. In other words, we have to show that $\max(0, q - p) \leq q \leq \min(1, q + (1 - p))$. This is immediate since on the left-hand side we decrease q by some non-negative amount while on the right-hand side we are increasing it by a non-negative amount.

Proof of Lemma 2 That D3$''$ implies D3$'$ is trivial. Therefore let us assume that $\mathcal{D}$ satisfies D3$'$ and prove that D3$''$ holds.

Consider $f, g \in \mathcal{D}$ such that $f + g \in \mathcal{L}_1$. By D4$'$ we have that $\frac{1}{2}f, \frac{1}{2}g \in \mathcal{D}$. We can always write $f = p - q, g = r - s$ for some $p, q, r, s \in \mathcal{H}$. It follows that

$$\left(\frac{1}{2}p + \frac{1}{2}r\right) - \left(\frac{1}{2}q + \frac{1}{2}r\right) \in \mathcal{D},$$
$$\left(\frac{1}{2}q + \frac{1}{2}r\right) - \left(\frac{1}{2}q + \frac{1}{2}s\right) \in \mathcal{D}.$$

Using D3$'$ we obtain that $(\frac{1}{2}p + \frac{1}{2}r) - (\frac{1}{2}q + \frac{1}{2}s) = \frac{1}{2}f + \frac{1}{2}g \in \mathcal{D}$. Using D4$'$ again, we have the thesis.

Proof of Theorem 2 Let us prove the five points of the theorem.

1. Assume that $\mathcal{D}$ satisfies D1$'$–D4$'$ and show that $\mathcal{R}_\mathcal{D}$ is coherent.
   - D1. Consider $f \in \mathcal{L}^+$ and let $g := f / \sup|f|$, so that $g \in \mathcal{L}_1^+$. As a consequence of D1$'$, $g \in \mathcal{D}$, whence $f \in \mathcal{R}_\mathcal{D}$.
   - D2. Trivial.
   - D3. Take $f, g \in \mathcal{R}_\mathcal{D}$, so that there are $f', g' \in \mathcal{D}, \lambda', \lambda'' > 0$ with $f = \lambda' f', g = \lambda'' g'$. If we let $\lambda := \lambda' + \lambda''$, we have that $f/(2\lambda), g/(2\lambda) \in \mathcal{D}$ by D4$'$ and at the same time $(f + g)/(2\lambda) \in \mathcal{L}_1$, so that $(f + g)/(2\lambda) \in \mathcal{D}$ by D3$''$. As a consequence, $f + g \in \mathcal{R}_\mathcal{D}$.
   - D4. Trivial.
2. That $\mathcal{D}_\mathcal{R}$ satisfies D1$'$–D4$'$ once $\mathcal{R}$ is coherent is straightforward.

3. We need to show that, under coherence of $\mathcal{R}, \mathcal{D}$, it holds that $\mathcal{R}_{\mathcal{D}_{\mathcal{R}}} = \mathcal{R}$ and $\mathcal{D}_{\mathcal{R}_{\mathcal{D}}} = \mathcal{D}$. For the first equality, that $\mathcal{R}_{\mathcal{D}_{\mathcal{R}}} \supseteq \mathcal{R}$ follows from

$$\mathcal{R}_{\mathcal{D}_{\mathcal{R}}} = \{\lambda g : \lambda > 0, g \in \mathcal{D}_{\mathcal{R}}\} = \left\{\lambda \frac{g}{\sup |g|} : \lambda > 0, g \in \mathcal{R}\right\} \supseteq \mathcal{R}$$

by taking $\lambda := \sup |g|$.
To show that $\mathcal{R}_{\mathcal{D}_{\mathcal{R}}} \subseteq \mathcal{R}$, consider any h in $\mathcal{R}$, implying that $\frac{h}{\sup |h|} \in \mathcal{D}_{\mathcal{R}}$. Note that $\sup |h| > 0$; indeed, if $\sup |h|$ would be 0 then $h = 0$. For any given $\lambda > 0$, this implies that $\lambda \frac{h}{\sup |h|} = \lambda' h$ for $\lambda' := \frac{\lambda}{\sup |h|} > 0$ indeed belongs to $\mathcal{R}$, because this set satisfies D4.
Let us prove now the equality $\mathcal{D}_{\mathcal{R}_{\mathcal{D}}} = \mathcal{D}$. Take $\mathcal{D}$ satisfying D1′–D4′. To see that $\mathcal{D}_{\mathcal{R}_{\mathcal{D}}} \subseteq \mathcal{D}$, take $f \in \mathcal{D}_{\mathcal{R}_{\mathcal{D}}}$. Then there is some $g \in \mathcal{R}_{\mathcal{D}}$ such that $f = \frac{g}{\sup |g|}$. Since $g \in \mathcal{R}_{\mathcal{D}}$, there is some $h \in \mathcal{D}$ and some $\lambda > 0$ such that $g = \lambda h$, whence $f = \frac{\lambda h}{\sup |\lambda h|} = \frac{h}{\sup |h|}$. Since $f \in \mathcal{L}_1$, we deduce from D4′ that $f \in \mathcal{D}$.
Conversely, take $f \in \mathcal{D}$. Then $f \in \mathcal{R}_{\mathcal{D}}$ and as a consequence $\frac{f}{\sup |f|} \in \mathcal{D}_{\mathcal{R}_{\mathcal{D}}}$. Since $\mathcal{D}_{\mathcal{R}_{\mathcal{D}}}$ satisfies D4′ and $f \in \mathcal{L}_1$, we deduce that $f \in \mathcal{D}_{\mathcal{R}_{\mathcal{D}}}$.
4. Since $\mathcal{D} \subseteq \mathcal{R}_{\mathcal{D}}$ and $\mathcal{R}_{\mathcal{D}}$ is coherent (apply 1), then $\mathcal{D}$ avoids partial loss by (Miranda and Zaffalon 2010, Proposition 3(e)). This implies that the minimal coherent extension of $\mathcal{D}$ exists. Let us call it $\mathcal{E}$. That $\mathcal{R}_{\mathcal{D}} \subseteq \mathcal{E}$ is trivial. Since $\mathcal{R}_{\mathcal{D}}$ is coherent, then by definition of $\mathcal{E}$, it must be that $\mathcal{R}_{\mathcal{D}} \supseteq \mathcal{E}$. As a consequence $\mathcal{R}_{\mathcal{D}}$ is the natural extension.
Let us show that $\mathcal{R}_{\mathcal{D}} \cap \mathcal{L}_1 \subseteq \mathcal{D}$ (the converse inclusion is trivial). Consider $f \in \mathcal{R}_{\mathcal{D}} \cap \mathcal{L}_1$. Then $f = \lambda g$ for some $\lambda > 0, g \in \mathcal{D}$. And since $\lambda g \in \mathcal{L}_1$, we get by D4′ that $f = \lambda g \in \mathcal{D}$. This shows that $\mathcal{D}$ is coherent relative to $\mathcal{L}_1$.
5. ($\Rightarrow$) Showing that $\mathcal{D}_{\mathcal{R}}$ satisfies D0′ is equivalent to showing that the preference relation it induces is weakly Archimedean, thanks to Theorem 1. Therefore consider $g \in \mathcal{D}_{\mathcal{R}} \setminus \mathcal{L}_1^+$ such that $g = p - q$ for some $p, q \in \mathcal{H}$. We want to show that there is $\alpha \in (0, 1)$ such that $\alpha p - q \in \mathcal{D}_{\mathcal{R}}$.
Thanks to point 4, we have that $g \in \mathcal{R} \setminus \mathcal{L}^+$ and hence by strict desirability that there is $\delta > 0$ such that $g - \delta \in \mathcal{R}$. Choose $\alpha \in (0, 1)$ so that $(1 - \alpha)p \leq \delta$. Then $\alpha p - q \geq p - q - \delta = g - \delta$, whence $\alpha p - q \in \mathcal{D}_{\mathcal{R}}$.
($\Leftarrow$) Finally, let us assume that $\mathcal{D}_{\mathcal{R}}$ satisfies D0′–D4′ and show that for all $h \in \mathcal{R} \setminus \mathcal{L}^+$ and $k \in \mathcal{L}$, there is $\alpha \in (0, 1)$ such that $\alpha h + (1 - \alpha)k \in \mathcal{R}$. If we prove this, then it is enough to take a constant $k < 0$ to get that $h - \delta \in \mathcal{R}$, with $\delta := -\frac{1-\alpha}{\alpha} k > 0$. This means that $\mathcal{R}$ would be strictly desirable.
Now then consider the positive and negative parts of k, namely, k^+, k^-, such that $k = k^+ - k^-$. Note that for the thesis it is enough to show that there is $\alpha \in (0, 1)$ such that $\alpha \frac{h}{2} - (1 - \alpha)k^- \in \mathcal{R}$, given that $\alpha \frac{h}{2} + (1 - \alpha)k^+$ belongs to $\mathcal{R}$ already, whence by additivity the result follows.
To that end, choose $\lambda > 0$ such that $f := \frac{h}{2\lambda}$ and $g := \frac{k^-}{\lambda}$ satisfy the inequalities in D0′. Then there is $\alpha \in (0, 1)$ such that $\alpha \frac{h}{2\lambda} - (1 - \alpha)\frac{k^-}{\lambda} \in \mathcal{D}_{\mathcal{R}}$; and as a consequence, $\alpha \frac{h}{2} - (1 - \alpha)k^- \in \mathcal{R}$.

Proof of Theorem 3 Thanks to Theorems 1 and 2, we have an equivalence between coherent preference relations $\succ$ and sets of desirable gambles coherent relative to $\mathcal{L}_1$; the latter are in a one-to-one correspondence with their natural extensions, from which we deduce the first part of the theorem. The second part is granted again by Theorems 1 and 2.

Proof of Proposition 3 Assume first of all that $\succ$ is negatively transitive, and let $f, g \notin \mathcal{R}$. Then $f \not\succ 0$ and $g \not\succ 0$, from which it follows that $0 \not\succ -g$. Applying the negative transitivity of $\succ$, we deduce that $f \not\succ -g$, or, equivalently, that $f + g \not\succ 0$, whence $f + g \notin \mathcal{R}$.

Conversely, assume that Eq. (7) holds and let $p, q, r \in \mathcal{H}$ satisfy $p \not\succ q$ and $q \not\succ r$. Then $p - q \notin \mathcal{R}$ and $q - r \notin \mathcal{R}$, whence $p - r \notin \mathcal{R}$, and using the correspondence between $\succ$ and $\mathcal{R}$ we conclude that $p \not\succ r$.

Proof of Proposition 4 From (10) and (11), we see that simply defining

$$p_2 := \psi(p_1) \tag{13}$$

for all horse lotteries $p_1 \in \mathcal{H}_z$, we obtain a correspondence between $\mathcal{H}$ and $\mathcal{H}_z$ made of elements that are equally valuable for us: their respective probability currencies are just the same. Moreover, the correspondence is one-to-one as it follows from Zaffalon and Miranda (2017, Remark 5 in p. 1098) (note also that ψ is a linear operator).

It is now a simple step to prove the two inclusions.

- $\mathcal{R}_2 \subseteq \mathcal{R}_1$ If $\lambda(p_2 - q_2) \in \mathcal{R}_2$ for some $p_2, q_2 \in \mathcal{H}$ and $\lambda > 0$, then also $\lambda\psi(p_1 - q_1) \in \mathcal{R}_1$, where p_1, q_1 are obtained through (13). This follows from the identity $p_2 - q_2 = \psi(p_1 - q_1)$, as obtained from linearity of ψ, and the fact that the lotteries defined through (13) are equally valuable.
- $\mathcal{R}_1 \subseteq \mathcal{R}_2$ Similarly, consider any $\lambda\psi(p_1 - q_1) \in \mathcal{R}_1$ and let $p_2 := \psi(p_1), q_2 := \psi(q_1)$, so that $p_2, q_2 \in \mathcal{H}$. It follows that $\lambda(p_2 - q_2) \in \mathcal{R}_2$ for the same reasons given in the proof of the converse inclusion.

Proof of Proposition 5 It follows from (Miranda et al. 2012, Proposition 29) that $\hat{\mathcal{R}}$ is the smallest conglomerable and coherent set of gambles that includes $\mathcal{R}_\Omega$ and $\mathcal{R}|\{\omega\}$ (for all $\omega \in \Omega$). It includes $\mathcal{R}|\Omega$ by construction. We are left to show that $\hat{\mathcal{R}}$ induces both $\mathcal{R}_\Omega$ and $\mathcal{R}_\mathcal{X}$.

We begin by proving that the Ω-marginal of $\hat{\mathcal{R}}$ is $\mathcal{R}_\Omega$. Consider an Ω-measurable gamble $f \in \hat{\mathcal{R}}$. Then there are $g \in \mathcal{R}_\Omega$ and $h \in \mathcal{R}|\Omega$ such that $f \geq g + h$. For any $\omega \in \Omega$, it holds that

$$0 \leq \sup_{x \in \mathcal{X}} h(\omega, x) \leq \sup_{x \in \mathcal{X}} (f(\omega, x) - g(\omega, x)) = f'(\omega) - g'(\omega),$$

where in last equality we are using that both f, g are Ω-measurable, and are denoting by f', g' their equivalent representations as gambles on Ω.

Thus, $f \geq g$, and since $\mathcal{R}_\Omega$ is a coherent set of gambles we conclude that also $f \in \mathcal{R}_\Omega$. The converse inclusion follows from Eq. (12).

Next, consider an $\mathcal{X}$-measurable gamble $f \in \hat{\mathcal{R}}$. Then there are $g \in \mathcal{R}_\Omega$ and $h \in \mathcal{R}|\Omega$ such that $f \geq g+h$. If $g \neq 0$, then there exists $\omega \in \Omega$ such that $g'(\omega) > 0$. Then,

$$fI_{\{\omega\}} \geq g'(\omega) + hI_{\{\omega\}} \geq hI_{\{\omega\}},$$

and since $h(\omega, \cdot) \in \mathcal{R}_\mathcal{X} \cup \{0\}$, either $h(\omega, \cdot) \in \mathcal{R}_\mathcal{X}$, in which case $f(\omega, \cdot) \in \mathcal{R}_\mathcal{X}$, or $h(\omega, \cdot) = 0$, whence $f(\omega, \cdot) \geq g'(\omega) > 0$, meaning that also $f(\omega, \cdot) \in \mathcal{R}_\mathcal{X}$. Finally, the proof when $g = 0$ is similar (in that case we can pick any $\omega \in \Omega$ such that $hI_{\{\omega\}} \neq 0$, since one is bound to exist). Again, the converse inclusion follows from Eq. (12).

Proof of Proposition 6 The result follows immediately from Definition 31, taking into account that $\underline{P}'_\Omega(\underline{P}(\cdot|\Omega))$ is a coherent lower prevision by Walley (1991, Section 6.7.2).

Proof of Proposition 7 We make a circular proof.

- $(i) \Rightarrow (ii)$ It $f \notin \mathcal{R}$, it follows from Eq. (3) that $\underline{P}(f) = \overline{P}(f) \leq 0$, whence $\underline{P}(-f) = -\overline{P}(f) \geq 0$, and therefore $\underline{P}(\varepsilon - f) > 0$ for every $\varepsilon > 0$. Thus, Eq. (3) implies that $\varepsilon - f \in \mathcal{R}$.
- $(ii) \Rightarrow (iii)$ Assume ex-absurdo the existence of $f, g \notin \mathcal{R}$ such that $f+g-\varepsilon \in \mathcal{R}$. It follows from (ii) that $\frac{\varepsilon}{4} - f, \frac{\varepsilon}{4} - g \in \mathcal{R}$, and applying D3 we conclude that $-\frac{\varepsilon}{2} \in \mathcal{R}$, a contradiction with D2.
- $(iii) \Rightarrow (i)$ If $\underline{P}$ is not linear, we can find a gamble f such that $\underline{P}(f) < \overline{P}(f)$. The conjugacy of the lower and upper previsions and Eq. (1) implies that for every $\varepsilon > 0$ the gambles $f - \underline{P}(f) - \varepsilon$ and $\overline{P}(f) - f - \varepsilon$ do not belong to $\mathcal{R}$, and (iii) implies then that for every $\delta > 0$ the gamble $(f - \underline{P}(f) - \varepsilon) + (\overline{P}(f) - f - \varepsilon) - \delta$ does not belong to $\mathcal{R}$. But for ε, δ satisfying $2\varepsilon + \delta < \overline{P}(f) - \underline{P}(f)$ this sum is a non-negative gamble. This is a contradiction with D1.

Proof of Proposition 8 This is an immediate consequence of Proposition 7.

References

Allais, M. (1953). Le comportement de l'homme rationnel devant le risque: Critique des postulats et axiomes de l'école Américaine. *Econometrica*, *21*(4), 503–546.

Anscombe, F. J., & Aumann, R. J. (1963). A definition of subjective probability. *The Annals of Mathematical Statistics*, *34*, 199–2005.

Armstrong, T. (1990). Conglomerability of probability measures on Boolean algebras. *Journal of Mathematical Analysis and Applications*, *150*(2), 335–358.

Augustin, T., Coolen, F., de Cooman, G., & Troffaes, M. (Eds.). (2014). *Introduction to imprecise probabilities*. Hoboken: Wiley.

Benavoli, A., Facchini, A., & Zaffalon, M. (2016a). Quantum mechanics: The Bayesian theory generalized to the space of Hermitian matrices. *Physical Review A*, *94*, 042106.

Benavoli, A., Facchini, A., & Zaffalon, M. (2016b). Quantum rational preferences and desirability. In *Proceedings of The 1st International Workshop on "Imperfect Decision Makers: Admitting Real-World Rationality" (NIPS 2016)*, JMLR.org, PMLR (Vol. 58, pp. 87–96).

Berti, P., Miranda, E., & Rigo, P. (2017). Basic ideas underlying conglomerability and disintegrability. *International Journal of Approximate Reasoning*, *88*, 387–400.

Bewley, T. (2002). Knightian decision theory: Part I. *Decisions in Economics and Finance*, *25*, 79–110. (reprint of Discussion Paper 807, Cowles Foundation, (1986)).

Carnap, R. (1980). *The problem of a more general concept of regularity* (Vol. 2, pp. 145–155). Berkeley: University of California Press.

Cerreira-Vioglio, S., Maccheroni, F., & Marinacci, M. (2016). Stochastic dominance analysis without the independence axiom. *Management Science*, *63*(4), 1097–1109.

Coletti, G., & Scozzafava, R. (2002). *Probabilistic logic in a coherent setting*. Dordrecht: Kluwer.

Corani, G., & Zaffalon, M. (2008). Learning reliable classifiers from small or incomplete data sets: The naive credal classifier 2. *Journal of Machine Learning Research*, *9*, 581–621.

Couso, I., & Moral, S. (2011). Sets of desirable gambles: Conditioning, representation, and precise probabilities. *International Journal of Approximate Reasoning*, *52*, 1034–1055.

Cozman, F. G. (2012). Sets of probability distributions, independence, and convexity. *Synthese*, *186*(2), 577–600.

De Bock, J. (2017). Independent natural extension for sets of desirable gambles: Williams-coherence to the rescue. In A. Antonucci, G. Corani, I. Couso, & S. Destercke (Eds.), *ISIPTA'17: Proceedings of the 10th International Symposium on Imprecise Probability: Theories and Applications* (Vol. 62, pp. 121–132). PMLR.

De Bock, J., & de Cooman, G. (2014). An efficient algorithm for estimating state sequences in imprecise hidden Markov models. *Journal of Artificial Intelligence Research*, *50*, 189–233.

De Bock, J., & de Cooman, G. (2015a). Conditioning, updating and lower probability zero. *International Journal of Approximate Reasoning*, *67*, 1–36.

De Bock, J., & de Cooman, G. (2015b). Credal networks under epistemic irrelevance: The sets of desirable gambles approach. *International Journal of Approximate Reasoning*, *56*, 178–207.

De Cooman, G., & Hermans, F. (2008). Imprecise probability trees: Bridging two theories of imprecise probability. *Artificial Intelligence*, *172*(11), 1400–1427.

De Cooman, G., & Miranda, E. (2009). Forward irrelevance. *Journal of Statistical Planning and Inference*, *139*(2), 256–276.

De Cooman, G., & Miranda, E. (2012). Irrelevance and independence for sets of desirable gambles. *Journal of Artificial Intelligence Research*, *45*, 601–640.

De Cooman, G., & Quaeghebeur, E. (2012). Exchangeability and sets of desirable gambles. *International Journal of Approximate Reasoning*, *53*, 363–395. (special issue in honour of Henry E. Kyburg, Jr.).

De Finetti, B. (1931). Sul significato soggettivo della probabilità. *Fundamenta Mathematicae*, *17*, 298–329.

De Finetti, B. (1937). La prévision: ses lois logiques, ses sources subjectives. Annales de l'Institut Henri Poincaré 7:1–68, English translation in Kyburg Jr. and Smokler (1964).

De Finetti, B. (1970). *Teoria delle Probabilità*. Turin: Einaudi.

De Finetti, B. (1974). Theory of probability: A critical introductory treatment (Vol. 1). Chichester: Wiley, English translation of de Finetti (1970).

Del Amo, A. G., & Ríos Insua, D. (2002). A note on an open problem in the foundations of statistics. *RACSAM*, *96*(1), 55–61.

DiBella, N. (2018). The qualitative paradox of non-conglomerability. *Synthese*, *195*(3), 1181–1210.

Dietterich, T. G. (2017). Steps toward robust artificial intelligence. *AI Magazine*, *38*(3), 3–24.

Domshlak, C., Hüllermeier, E., Kaci, S., & Prade, H. (2011). Preferences in AI: An overview. *Artificial Intelligence*, *175*(7–8), 1037–1052.

Dubins, L. E. (1975). Finitely additive conditional probabilities, conglomerability and disintegrations. *The Annals of Probability*, *3*, 88–99.

Dubois, D., Fargier, H., & Perny, P. (2003). Qualitative decision theory with preference relations and comparative uncertainty: An axiomatic approach. *Artificial Intelligence*, *148*(1–2), 219–260.

Dubra, J., Maccheroni, F., & Ok, E. (2004). Expected utility theory without the completeness axiom. *Journal of Economic Theory*, *115*, 118–133.

Galaabaatar, T., & Karni, E. (2013). Subjective expected utility with incomplete preferences. *Econometrica*, *81*(1), 255–284.

Ghirardato, P., & Siniscalchi, M. (2012). Ambiguity in the small and in the large. *Econometrica*, *80*(6), 2827–2847.

Hacking, I. (1967). Slightly more realistic personal probability. *Philosophy of Science*, *34*(4), 311–325.

Hájek, A., & Smithson, M. (2012). Rationality and indeterminate probabilities. *Synthese*, *187*(1), 33–48.

Hill, B., & Lane, D. (1985). Conglomerability and countable additivity. *Sankhya*, *47*, 366–379.

Howson, C. (2009). Can logic be combined with probability? Probably. *Journal of Applied Logic*, *7*, 177–187.
Howson, C. (2011). Bayesianism as a pure logic of inference. In P. S. Bandyopadhyay & M. R. Forster (Eds.), *Handbook of the philosophy of science, vol 7: Philosophy of statistics* (pp. 441–471). Amsterdam: Elsevier.
Kikuti, D., Cozman, F. G., & Filho, R. S. (2011). Sequential decision making with partially ordered preferences. *Artificial Intelligence*, *175*(7–8), 1346–1365.
Kyburg, H.E. Jr., & Smokler, H.E. (Eds.) (1964) *Studies in subjective probability* (2 Ed.). New York: Wiley (with new material) (1980).
Luce, R. D., & Krantz, D. H. (1971). Conditional expected utility. *Econometrica*, *39*(2), 253–271.
Machina, M. (1982). "Expected utility" analysis without the independence axiom. *Econometrica*, *50*(2), 277–323.
Miranda, E. (2008). A survey of the theory of coherent lower previsions. *International Journal of Approximate Reasoning*, *48*(2), 628–658.
Miranda, E., & de Cooman, G. (2007). Marginal extension in the theory of coherent lower previsions. *International Journal of Approximate Reasoning*, *46*(1), 188–225.
Miranda, E., & Zaffalon, M. (2010). Notes on desirability and conditional lower previsions. *Annals of Mathematics and Artificial Intelligence*, *60*(3–4), 251–309.
Miranda, E., & Zaffalon, M. (2015a). Independent products in infinite spaces. *Journal of Mathematical Analysis and Applications*, *425*(1), 460–488.
Miranda, E., & Zaffalon, M. (2015b). On the problem of computing the conglomerable natual extension. *International Journal of Approximate Reasoning*, *56*, 1–27.
Miranda, E., & Zaffalon, M. (2016). Conformity and independence with coherent lower previsions. *International Journal of Approximate Reasoning*, *78*, 125–137.
Miranda, E., Zaffalon, M., & de Cooman, G. (2012). Conglomerable natural extension. *International Journal of Approximate Reasoning*, *53*(8), 1200–1227.
Moral, S. (2005). Epistemic irrelevance on sets of desirable gambles. *Annals of Mathematics and Artificial Intelligence*, *45*, 197–214.
Nau, R. (2006). The shape of incomplete preferences. *The Annals of Statistics*, *34*(5), 2430–2448.
Nau, R. (2011). Risk, ambiguity, and state-preference theory. *Economic Theory*, *48*(2–3), 437–467.
Ok, E., Ortoleva, P., & Riella, G. (2012). Incomplete preferences under uncertainty: Indecisiveness in beliefs versus tastes. *Econometrica*, *80*(4), 1791–1808.
Pedersen, A. P. (2014). Comparative expectations. *Studia Logica*, *102*(4), 811–848.
Pigozzi, G., Tsoukiàs, A., & Viappiani, P. (2016). Preferences in artificial intelligence. *Annals of Mathematics and Artificial Intelligence*, *77*(3–4), 361–401.
Savage, L. J. (1954). *The foundations of statistics*. New York: Wiley.
Seidenfeld, T., Schervish, M. J., & Kadane, J. B. (1995). A representation of partially ordered preferences. *The Annals of Statistics*, *23*, 2168–2217. (reprinted in Seidenfeld et al. (1999), (pp. 69–129)).
Seidenfeld, T., Schervisch, M. J., & Kadane, J. B. (1998). Non-conglomerability for finite-valued finitely additive probability. *Sankhya*, *60*(3), 476–491.
Seidenfeld, T., Schervish, M. J., & Kadane, J. B. (1999). *Rethinking the foundations of statistics*. Cambridge: Cambridge University Press.
Seidenfeld, T., Schervisch, M. J., & Kadane, J. B. (2010). Coherent choice functions under uncertainty. *Synthese*, *172*(1), 157–176.
Shafer, G., Gillett, P. R., & Scherl, R. B. (2003). A new understanding of subjective probability and its generalization to lower and upper prevision. *International Journal of Approximate Reasoning*, *33*, 1–49.
Shimony, A. (1955). Coherence and the axioms of confirmation. *The Journal of Symbolic Logic*, *20*(1), 1–28.
Skyrms, B. (1980). *Causal necessity*. New Haven: Yale University Press.
Tversky, A., & Kahneman, D. (1992). Advances in prospect theory: Cumulative representation of uncertainty. *Journal of Risk and Uncertainty*, *5*(4), 297–323.
Van Camp, A. (2018). *Choice functions as a tool to model uncertainty*. Ph.D. thesis, University of Ghent.
Van Camp, A., de Cooman, G., Miranda, E., & Quaeghebeur, E. (2018a). Coherent choice functions, desirability and indifference. *Fuzzy Sets and Systems*, *341*(C), 1–36.
Van Camp, A., Miranda, E., & de Cooman, G. (2018b). Lexicographic choice functions. *International Journal of Approximate Reasoning*, *92*, 97–119.

Von Neumann, J., & Morgestern, O. (1947). *Theory of games and economic behaviour*. Princeton: Princeton University Press.
Walley, P. (1991). *Statistical reasoning with imprecise probabilities*. London: Chapman and Hall.
Williams, P. M. (1975). *Notes on conditional previsions*. Tech. rep., School of Mathematical and Physical Science, University of Sussex, UK, reprinted in Williams (2007).
Williams, P. M. (2007). Notes on conditional previsions. *International Journal of Approximate Reasoning*, *44*, 366–383. (revised journal version of Williams (1975)).
Zaffalon, M., & Miranda, E. (2013). Probability and time. *Artificial Intelligence*, *198*(1), 1–51.
Zaffalon, M., & Miranda, E. (2015). Desirability and the birth of incomplete preferences. CoRR abs/1506.00529. arxiv:1506.00529.
Zaffalon, M., & Miranda, E. (2017). Axiomatising incomplete preferences through sets of desirable gambles. *Journal of Artificial Intelligence Research*, *60*, 1057–1126.

Synthese (2021) 198 (Suppl 27):S6571–S6597
https://doi.org/10.1007/s11229-019-02512-2

S.I.: DECTHEORY&FUTOFAI

Subjective causal networks and indeterminate suppositional credences

Jiji Zhang[1] · **Teddy Seidenfeld**[2] · **Hailin Liu**[3]

Received: 24 February 2018 / Accepted: 9 December 2019 / Published online: 17 December 2019

Abstract

This paper has two main parts. In the first part, we motivate a kind of indeterminate, suppositional credences by discussing the prospect for a subjective interpretation of a causal Bayesian network (CBN), an important tool for causal reasoning in artificial intelligence. A CBN consists of a causal graph and a collection of interventional probabilities. The subjective interpretation in question would take the causal graph in a CBN to represent the causal structure that is believed by an agent, and interventional probabilities in a CBN to represent suppositional credences. We review a difficulty noted in the literature with such an interpretation, and suggest that a natural way to address the challenge is to go for a generalization of CBN that allows indeterminate credences. In the second part, we develop a decision-theoretic foundation for such indeterminate suppositional credences, by generalizing a theory of coherent choice functions to accommodate some form of act-state dependence. The upshot is a decision-theoretic framework that is not only rich enough to, so to speak, ground the probabilities in a subjectively interpreted causal network, but also interesting in its own right, in that it accommodates both act-state dependence and imprecise probabilities.

Keywords Causal Bayesian networks · Causal credal networks · Choice functions · Horse lotteries · Indeterminate credences · Interventions

Chapter 8 was originally published as Zhang, J., Seidenfeld, T. & Liu, H. Synthese (2021) 198 (Suppl 27): S6571–S6597. https://doi.org/10.1007/s11229-019-02512-2.

✉ Jiji Zhang
jijizhang@ln.edu.hk

Teddy Seidenfeld
teddy@stat.cmu.edu

Hailin Liu
sysulhl@aliyun.com

1 Department of Philosophy, Lingnan University, Hong Kong, China

2 Department of Philosophy, Carnegie Mellon University, Pittsburgh, USA

3 Institute of Logic and Cognition, Department of Philosophy, Sun Yat-sen University, Guangzhou, China

1 Introduction

The main purpose of this paper is twofold and will be pursued in two parts. In the first part, we aim to motivate a kind of indeterminate, suppositional credences by discussing the prospect for a subjective interpretation of a popular framework known as causal Bayesian network. The machinery of Bayesian network (Pearl 1988) has proved to be one of the most useful tools for knowledge representation and probabilistic reasoning in the field of artificial intelligence. Its importance has been further amplified in recent decades by the development of a causal semantics of the formalism, leading to breakthroughs in automated causal reasoning (Spirtes et al. 2000; Pearl 2009) and inspiring new insights in philosophical and psychological theorization about causation and causal knowledge (Glymour 2001; Woodward 2003; Gopnik and Schulz 2007; Danks 2014). Formally, a Bayesian network consists of a directed acyclic graph and a joint probability distribution over a set of random variables (where the graph and the probability are required to satisfy a Markov condition), and a *causal* Bayesian network is further associated with a number of interventional probability distributions. The term 'Bayesian' in the name may have suggested that the probabilities in the formalism are subjectively interpreted; indeed, Bayesian networks are also known as belief networks, and when used purely as a tool for probabilistic reasoning, the probabilities therein are often treated, at least implicitly, as experts' degrees of belief. By contrast, *causal* Bayesian networks are usually employed with an objective interpretation, in which probabilities are taken to represent (features of) an objective data generating process rather than subjective credences. However, it also seems natural, especially in psychological and decision-theoretic applications, to adopt a subjective interpretation of a causal Bayesian network, where probabilities therein are supposed to represent an agent's credences (and the graph is supposed to represent the causal structure believed by the agent). Subjective interpretations of this kind have been considered by a number of authors (Meek and Glymour 1994; Spirtes et al. 2000; Spohn 2000; Williamson 2001; Glymour and Danks 2007; Stern 2017).

As we will argue in this paper, once we intend causal Bayesian networks to model credences (and subjectively believed causal structures), there is a compelling reason to allow credences to be indeterminate and to be represented by sets of probability functions, in addition to the stock arguments in decision theory and formal epistemology in favour of non-sharp credences. Moreover, once sets of probability functions are in, the same consideration also implies that these sets are usually non-convex rather than convex. For this reason, in the second part where we attempt to provide a decision-theoretic foundation for such credences, we will build on Seidenfeld et al.'s (2010) work on choice functions; to our knowledge, their theory of coherent choice functions is currently the only decision theory that permits a non-convex set of probabilities to be uniquely characterized by choice patterns.

The second part of the paper is thus devoted to generalizing Seidenfeld et al.'s theory so as to provide a decision-theoretic foundation for the credences modeled by a causal Bayesian network (or credal network, as it is known in the literature when the probabilities involved are imprecise). Seidenfeld et al.'s theory does not yet meet our demand for the following reason. A causal Bayesian network does not involve just one probability distribution, but a number of *interventional* probability distributions.

Understood as credences, these interventional probabilities presumably represent credences under suppositions—for example, degrees of belief concerning a variable Y (say, whether the floor is wet) taking a certain value (say, 'yes'), under the supposition that a variable X (say, whether the sprinkler is on) were intervened to take a certain value (say, 'no') (Glymour and Danks 2007). If we think of hypothetical interventions as contemplated actions, as it is natural in this setting, then it is easy to construct choice situations on which these suppositional credences bear. However, such choice situations will in general exhibit act-state dependence, for interventional probability distributions are in general different given different interventions. Seidenfeld et al.'s theory assumes act-state independence, and we need to relax that assumption for our purpose.

The upshot is a decision-theoretic framework that is rich enough to, so to speak, ground the probabilities in a subjectively interpreted causal network. In addition, for readers with sympathies for an agency or decision-theoretic approach to causation (Price 1991; Menzies and Price 1993; Hitchcock 1996), our attempt in this paper may be appreciated as an effort to elucidate a notion of *imprecise agent probabilities* that an agency theory of causation might borrow if it were pressed to allow rational agents to have indeterminate credences. Moreover, from a purely decision-theoretic viewpoint, the generalized theory of coherent choice functions we are beginning to build in this paper is also interesting in its own right; to our knowledge, no theory is yet on offer that accommodates both act-state dependence and imprecise probabilities.

The rest of the paper will proceed as follows. The first part comprises Sect. 2, where we elaborate on the motivation for the kind of indeterminate, suppositional credences sketched above. In particular, we highlight a difficulty with subjective causal Bayesian networks pointed out by Meek and Glymour (1994), which we propose to address by invoking imprecise probabilities and moving to the more general framework of causal credal networks. We show how the standard interventionist definition of causal structure is readily generalized to the more general framework, which may be exploited to reduce qualitative causal beliefs to suppositional credences.

We then switch to decision theory in the second part. In preparation for our generalization of Seidenfeld et al.'s (2010) theory of choice functions, we first generalize, in Sect. 3, Anscombe and Aumann's (1963) theory of coherent preference rankings over 'horse lotteries', on which Seidenfeld et al.'s theory is built. The generalized theory can be used to characterize *determinate* suppositional credences. Finally, in Sect. 4, we outline a generalized theory of coherent choice functions that aims at *indeterminate* suppositional credences. We conclude in Sect. 5.

2 Credences in a subjective causal network

The generalized theory of coherent choice functions we will develop later is designed to capture a fairly complex kind of credences. We are driven to consider this kind in our attempt to construe causal Bayesian networks subjectively. In this section we describe in more detail our reasons for taking such credences seriously.

Fig. 1 A DAG over 4 variables

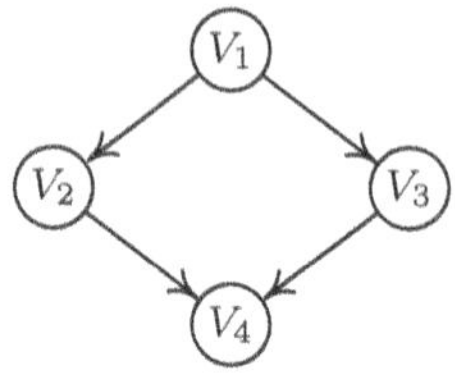

2.1 Causal Bayesian networks

A Bayesian network is defined over a set of random variables $\mathbf{V}$. A common practice is to assume $\mathbf{V}$ to be a set of categorical random variables whose ranges are finite (i.e., every variable in $\mathbf{V}$ has finitely many possible values), which we follow in this paper. A directed graph over $\mathbf{V}$ is a vertex-edge graph whose set of vertices stands in one-one correspondence with $\mathbf{V}$, and whose edges are ordered pairs of distinct vertices. As usual, we will not take pains to distinguish between a variable and the vertex that represents the variable, but simply treat vertices in a graph as variables. And we usually refer to an edge (X, Y) as an arrow from X to Y (written as $X \to Y$.) Given a directed graph $\mathcal{G}$ over $\mathbf{V}$, variable X is called a *parent* of variable Y in $\mathcal{G}$ just in case $\mathcal{G}$ contains an arrow from X to Y. A directed path from X to Y is a sequence of distinct vertices $\langle X = V_0, \ldots, V_n = Y \rangle$ such that V_{i-1} is V_i's parent for every $1 \leq i \leq n$. Y is said to be a *descendant* of X if $Y = X$ or there is a directed path from X to Y. Otherwise, Y is called a *non-descendant* of X. A *directed acyclic graph (DAG)* is a directed graph in which there is no directed path from any variable to any of its parents. A *Bayesian network (BN)* over $\mathbf{V}$ is formally a pair $(\mathcal{G}, P)$, where $\mathcal{G}$ is a DAG over $\mathbf{V}$ and P is a joint probability distribution of $\mathbf{V}$ such that they satisfy a Markov condition. There are several equivalent formulations of the Markov condition. The formulation that is often used in the philosophical literature is that it is true of P that every variable in $\mathbf{V}$ is probabilistically independent of its non-descendants in $\mathcal{G}$ conditional on its parents in $\mathcal{G}$. That is, $\mathcal{G}$ entails a number of conditional independence statements that P must satisfy in order to satisfy the Markov condition.

For present purposes, a more convenient (and equivalent) formulation of the Markov condition is that P can be factorized according to $\mathcal{G}$ as follows: $P(\mathbf{V}) = \prod_{X \in \mathbf{V}} P(X|\mathbf{Pa}_{\mathcal{G}}(X))$, where $\mathbf{Pa}_{\mathcal{G}}(X)$ denotes the set of parents of X in $\mathcal{G}$. When $(\mathcal{G}, P)$ satisfies the Markov condition, we also say that P is Markov to $\mathcal{G}$. For example, a probability distribution that is Markov to the DAG in Fig. 1 must satisfy that $P(V_1, V_2, V_3, V_4) = P(V_1)P(V_2|V_1)P(V_3|V_1)P(V_4|V_2, V_3)$.

BNs, especially those with sparse DAGs (DAGs with sufficiently few arrows), provide a powerful tool for probabilistic inference, for the factorization usually implies a significant reduction of the number of parameters needed to encode the joint probability distribution and facilitates computation of various marginal and conditional probabilities (Pearl 1988). Meanwhile, the DAG in a BN is often readily interpretable, as a representation of qualitative causal knowledge. Once the causal interpretation is made explicit and rigorous, the resulting framework, known as *causal Bayesian network (CBN)*, has been pivotal in the fruitful research on causality in artificial intelligence and related fields in the past few decades (Spirtes et al. 2000; Pearl 2009).

A CBN over $\mathbf{V}$ does not just model the *observational* (or *actual*) joint probability distribution of $\mathbf{V}$, but model a number of *interventional* (or *counterfactual*) distributions as well. The standard setup is that for every subset $\mathbf{S} \subseteq \mathbf{V}$ and every possible value configuration $\mathbf{s}$ for $\mathbf{S}$, there is an interventional distribution for $\mathbf{V} - \mathbf{S}$, given that $\mathbf{S}$ is (exogenously) intervened to take value $\mathbf{s}$. We denote such a probability distribution by $P^{do(\mathbf{S}=\mathbf{s})}$, using Pearl's (2009) celebrated do-operator. The observational distribution can be stipulated as the distribution under the null intervention, $P^{do(\emptyset)}$. For present purposes, a CBN can be formally defined as follows, generalizing the above definition of a BN in terms of factorization.

Definition 1 (*Causal Bayesian network*) A causal Bayesian network over $\mathbf{V}$ is a pair $(\mathcal{G}, \{P^{do(\mathbf{S}=\mathbf{s})}\})$, where $\mathcal{G}$ is a DAG over $\mathbf{V}$ and $\{P^{do(\mathbf{S}=\mathbf{s})}\}$ is a collection of interventional distributions that satisfy the following **intervention principle**: for every $\mathbf{S} \subseteq \mathbf{V}$ and every possible value configuration $\mathbf{s}$ for $\mathbf{S}$,

$$P^{do(\mathbf{S}=\mathbf{s})}(\mathbf{V} - \mathbf{S}) = \prod_{X \in \mathbf{V}-\mathbf{S}} P(X|\mathbf{Pa}_{\mathcal{G}}(X)),$$

(provided that the conditional probabilities on the right-hand side are all defined), where $P = P^{do(\emptyset)}$ is the observational distribution, and for each $P(X|\mathbf{Pa}_{\mathcal{G}}(X))$ on the right-hand side, every $Y \in \mathbf{S} \cap \mathbf{Pa}_{\mathcal{G}}(X)$ is assigned the value that is consistent with $\mathbf{S} = \mathbf{s}$.[1]

For example, for a CBN with the DAG in Fig. 1, if we consider the interventional distribution given $do(V_2 = v_2)$, it should factorize as follows: $P^{do(V_2=v_2)}(V_1, V_3, V_4) = P(V_1)P(V_3|V_1)P(V_4|V_2 = v_2, V_3)$. The key idea behind the intervention principle is that interventions in question are effective and precise—effective in the sense that they completely fix the values of the variables they directly target (e.g., V_2 in the above example), and precise in the sense that they do not affect the local mechanisms for the variables they do not directly target (e.g., V_1, V_3, and V_4 in the above example, so that $P(V_1)$, $P(V_3|V_1)$, and $P(V_4|V_2, V_3)$ remain invariant under the intervention).

The intervention principle implies the Markov condition (when $\mathbf{S} = \emptyset$), and may be regarded as a generalization of the Markov condition to all interventional contexts. The intervention principle (or the causal Markov condition) is plausible only if $\mathbf{V}$ is assumed to be *causally sufficient* in the sense that it does not leave out confounding variables (Spirtes et al. 2000). In this paper, we will not worry about causally insufficient systems; we assume that causal sufficiency is satisfied (or is believed to be satisfied) by $\mathbf{V}$.

[1] Note that we take $P^{do(\mathbf{S}=\mathbf{s})}$ to be a probability distribution over $\mathbf{V} - \mathbf{S}$ rather than over $\mathbf{V}$. A more common formulation is to take $P^{do(\mathbf{S}=\mathbf{s})}$ to be a distribution over $\mathbf{V}$ and take the intervention principle to require that

$$P^{do(\mathbf{S}=\mathbf{s})}(\mathbf{V}) = \begin{cases} \prod_{X \in \mathbf{V}-\mathbf{S}} P(X|\mathbf{Pa}_{\mathcal{G}}(X)) & \text{for values of } \mathbf{V} \text{ consistent with } \mathbf{S} = \mathbf{s}, \\ 0 & \text{for values of } \mathbf{V} \text{ inconsistent with } \mathbf{S} = \mathbf{s}. \end{cases}$$

Our formulation in the main text is entailed by and slightly weaker than this formulation. In particular, this formulation makes explicit that $P^{do(\mathbf{S}=\mathbf{s})}(\mathbf{S} = \mathbf{s}) = 1$ while ours does not. We thank an anonymous referee for this observation. We adopt this weaker formulation because it is more convenient for our subsequent discussions. All our points remain valid under the stronger formulation.

In a CBN, the DAG is usually referred to as the causal graph that represents the causal structure over $\mathbf{V}$. Arrows in a causal graph are often said to represent *direct* causal relevance relative to $\mathbf{V}$; that is, $X \rightarrow Y$ represents that X is directly causally relevant to Y relative to $\mathbf{V}$.

As it is usually done in the literature, we assume that the observational distribution is strictly positive, so that all the relevant conditional probabilities are defined. Under this assumption, we note two facts about the arrows in a CBN, which were first proved by Zhang and Spirtes (2011, Theorem 4).

Proposition 1 *Let* $(\mathcal{G}, \{P^{do(\mathbf{S}=\mathbf{s})}\})$ *be a causal Bayesian network over* $\mathbf{V}$. *For every* $X \neq Y \in \mathbf{V}$, *and* $\mathbf{Z} = \mathbf{V} - \{X, Y\}$, *if there exist two distinct values,* $x_1 \neq x_2$, *for* X *and a value configuration* $\mathbf{z}$ *for* $\mathbf{Z}$, *such that*

$$P^{do(X=x_1 \wedge \mathbf{Z}=\mathbf{z})}(Y) \neq P^{do(X=x_2 \wedge \mathbf{Z}=\mathbf{z})}(Y),$$

then there is an arrow from X *to* Y *in* $\mathcal{G}$.

In words, Proposition 1 says that if there exist interventions that set X to different values while keeping all variables other than X and Y fixed at some values, and that result in different probability distributions for Y, then X is a (causal) parent of Y. This statement is intuitively compelling; in fact, the antecedent of the statement is usually offered as a definition of direct causal relevance relative to $\mathbf{V}$, especially in an explicitly interventionist account of causation (Woodward 2003; Pearl 2009). It is interesting to see that the sufficiency of the definiens in this interventionist definition of direct causal relevance is a simple consequence of Definition 1 (provided that the observational distribution is strictly positive).[2]

The converse of Proposition 1 does not hold. It is easy to see that for every CBN $(\mathcal{G}, \{P^{do(\mathbf{S}=\mathbf{s})}\})$ over $\mathbf{V}$ and every DAG $\mathcal{G}'$ over $\mathbf{V}$ that is a supergraph of $\mathcal{G}$ (i.e., $\mathcal{G}'$ contains all the arrows in $\mathcal{G}$ and possibly more), $(\mathcal{G}', \{P^{do(\mathbf{S}=\mathbf{s})}\})$ is also a CBN (i.e., satisfies Definition 1, simply because for every $X \in \mathbf{V}$, $P(X|\mathbf{Pa}_{\mathcal{G}'}(X)) = P(X|\mathbf{Pa}_{\mathcal{G}}(X))$, thanks to the Markov condition.) Nonetheless, the (necessity of the definiens in the) interventionist definition of direct causal relevance can be motivated by appeal to the following fact:

Proposition 2 *Let* $(\mathcal{G}, \{P^{do(\mathbf{S}=\mathbf{s})}\})$ *be a causal Bayesian network over* $\mathbf{V}$, *and let* $\mathcal{G}_I$ *be the graph over* $\mathbf{V}$ *that contains exactly the arrows that satisfy the condition in Proposition 1. Then* $(\mathcal{G}_I, \{P^{do(\mathbf{S}=\mathbf{s})}\})$ *is also a causal Bayesian network over* $\mathbf{V}$.

[2] The positivity of the observational distribution is needed for Proposition 1, because the intervention principle in Definition 1 puts constraints on an interventional probability distribution only if the relevant observational conditional probabilities are defined. As an illustration, here is a simple kind of situation in which failure of positivity can easily induce a counterexample to Proposition 1. Consider three binary variables X, Y, Z, among which X and Z are exogenous, meaning that neither of them is affected by any intervention of the other two variables. Suppose Y is affected by X and Z in such a way that if $Z = 0$, then the value of X does not make a difference to Y, whereas if $Z = 1$, then the value of X does. In other words, $P^{do(X=0 \wedge Z=0)}(Y) = P^{do(X=1 \wedge Z=0)}(Y)$, while $P^{do(X=0 \wedge Z=1)}(Y) \neq P^{do(X=1 \wedge Z=1)}(Y)$. However, if the observational distribution $P^{do(\emptyset)}$ puts zero mass on $Z = 1$, then all that is needed to satisfy the intervention principle in Definition 1 is just a graph with an arrow from Z to Y but without an arrow from X to Y. In particular, given this graph, the intervention principle does not constrain $P^{do(X=0 \wedge Z=1)}(Y)$ or $P^{do(X=1 \wedge Z=1)}(Y)$, for $P^{do(\emptyset)}(Y|Z = 1)$ is undefined. So, despite the disequality $P^{do(X=0 \wedge Z=1)}(Y) \neq P^{do(X=1 \wedge Z=1)}(Y)$, the arrow from X to Y is not needed to meet Definition 1.

In other words, those arrows in a CBN that do not satisfy the interventionist definition of direct causal relevance are in a sense redundant: take them out and the network remains a CBN (with the same collection of interventional probability distributions). Together with Proposition 1, this entails that if a system (characterized by a set of variables) can be modeled by a CBN, then among such CBNs there is a unique simplest one, whose graph contains all and only those arrows that respect the interventionist definition of direct causal relevance. It is then natural to take this graph as the correct representation of the causal structure associated with the given collection of interventional distributions. To put it differently, if justification is needed for positing direct causal relevancies, then given a collection of interventional distributions, the intervention principle justifies positing those that satisfy the interventionist definition but does not justify positing more. Therefore, even though the definition of a CBN does not explicitly invoke the interventionist definition of direct causal relevance, this account emerges naturally in view of Propositions 1 and 2.

This provides a compelling reason to restrict attention to those CBNs in which the graph accurately represents the causal structure as defined by the interventionist account of direct causal relevance.[3] With this restriction, the pair that constitutes a CBN is reducible to its second element, the collection of interventional probability distributions, because the first element, the causal graph, is determined by the second element. This of course does not mean that the graph can be dispensed with in practice, but it suggests that in order to make sense of a subjective interpretation of a CBN, it is sufficient to make sense of a subjective interpretation of the interventional probabilities.

We stress this point because in a subjective interpretation of CBNs, it is natural to treat the graph in a CBN as representing the causal structure that is known or believed by an agent, and an immediate question is what it is for an agent to believe a causal structure, or believe that X is directly causally relevant to Y relative to $\mathbf{V}$ (since a causal structure in this context is just a set of statements about direct causal relevance). In light of the foregoing discussions, the rough answer we recommend is that it is reducible to or grounded in those interventional probabilities, which are interpreted as representing the agent's suppositional credences, e.g., credences concerning Y given the supposition that all the other variables are manipulated to take certain values. That is, what it is to believe a causal structure is to have a certain pattern of suppositional credences. A helpful reference here is the agency theory of causation (Price 1991; Menzies and Price 1993). According to this theory, what it is for an event C to cause another event E is to be elucidated in terms of the so-called 'agent probabilities', which are akin to what we call suppositional credences in this paper. Our suggestion is that instead of taking causation itself as the analysandum of an agency theory, one can try to analyze beliefs about causation along a similar line. It is not the purpose of this paper to develop a detailed account of qualitative causal belief, but this is the direction we would go if we

[3] A possible view to take here is that a CBN is individuated by its collection of interventional distributions, and so a mere change of the graph (e.g., to a proper supergraph) does not change the CBN. Then the simplest graph picked out by Propositions 1 and 2 is naturally regarded as the canonical form of the first element of a CBN. Under this view, the restriction is not a restriction on CBNs but rather a restriction to canonical forms. We thank an anonymous referee for raising this issue.

were pressed to make sense of our talk of beliefs in a causal structure. (Some further illustration of the idea is given in Sect. 2.4.)

2.2 A difficulty with subjective causal Bayesian networks

Immediately after they formulated a subjective reading of the causal Markov condition, Christopher Meek and Clark Glymour (1994) noted the following difficulty (attributed to Brian Skyrms): "[Nothing] prevents one from believing that an odd lot of dissimilar appearing coins of different manufacture and history all have the exact same bias." (p. 1006). The worry is that in such a case, the person's credences concerning outcomes of tossing these coins do not seem to satisfy the Markov condition with the causal structure the person believes. For concreteness, suppose we represent the outcome of one flip of a coin by X and that of a different coin by Y. In the intended scenario, the agent believes that X and Y are causally unconnected, without any causal relation between them, or any common cause (thus the two variables alone are causally sufficient). In other words, the believed causal structure is properly represented by a graph over X and Y alone without any arrow between them. Any joint probability distribution over X and Y that is Markov to this causal graph should factorize into a product of the marginal distribution of X and that of Y, implying that X and Y are probabilistically independent. However, since the agent somehow believes that the two coins have the same, unknown bias, if her credences concerning the two variables are represented by a single probability distribution, it is most reasonable to use a joint distribution that features a probabilistic dependence between X and Y and hence is not Markov to the believed causal structure.

To simplify the example for the sake of illustration, suppose the agent somehow believes that the unknown, common bias is either against heads so that the chance of landing heads for both coins is 0.4, or against tails so that the chance of landing heads for both coins is 0.6. Here we have two possible chance distributions over X and Y, each of which satisfies the statement that X and Y are probabilistically independent, meeting the Markov condition with respect to the causal graph in question. However, if the agent's credence function over X and Y is to be represented by a single probability distribution, the most plausible candidate in this case is some nontrivial, linear mixture of the two chance distributions. A well known result that can be traced back at least to Yule (1903) is that in such a case (where the two joint distributions imply different marginal distributions for X and for Y), a nontrivial, linear mixture necessarily results in a joint distribution that features probabilistic dependence between X and Y. This suggests that such a situation cannot be plausibly modeled by a subjective CBN.

It may be tempting to respond by disputing the claim that in such a situation the agent believes that X and Y are not causally connected whatsoever. Rather, the response goes, the belief that the two coins have the same bias betrays a commitment to a more complex causal structure, in which X and Y share a common cause, namely the common bias or whatever hidden variable that is responsible for the common bias (and thus X and Y alone are not causally sufficient). With this common cause structure, it is only natural and certainly no violation of the Markov condition to entertain a probabilistic dependence between X and Y.

This response begs the question against the original worry, for the very point of the challenge was that it is not impossible—not even irrational—for an agent to believe that two coins have the same bias even though they share no common causal ancestry. The response amounts to no more than denying flatly the possibility or at least the rationality of such an agent. In our view, such an agent is plainly possible. Whether she is rational is less clear, though we are inclined to agree with Meek and Glymour (and Skyrms) that it is difficult to see a good argument against her rationality. In the absence of such arguments, it seems a little extreme to assume or theorize the situation away.

Another quite radical reaction is to give up altogether the project of subjectively interpreting causal networks. In his version of an interventionist decision theory, for example, Stern (2017) required that probabilities *within* a causal network should be interpreted as objective chances rather than credences. Before we try harder, this prohibition against a subjective interpretation of the probabilities in causal networks seems to be premature.

In this paper, we would like to highlight a possible route to salvage the subjectively interpreted causal Markov condition in the above example, by relaxing the assumption that credences are to be represented by a single, precise probability function. If we allow indeterminate or non-sharp credences, to be represented by a set of probability functions, then in the above simplified example, it is most natural to represent the credences by the set consisting of the two joint probability distributions that the agent regard as possible candidates for the unknown chance distribution. Then there is a perfectly good sense—called 'independence in the selection' by Couso et al. (2000) or 'complete independence' by Cozman (2012) (attributed to Seidenfeld)[4]—in which the indeterminate credences satisfy the statement that X and Y are independent, in the sense that every probability distribution in the set implies the said independence. The subjective Markov condition is then satisfied.[5]

We hasten to add that this is not the only reason to permit indeterminate probabilities in a causal network. A number of stock arguments in favor of relaxing the requirement of determinacy or precision—e.g., that sharp credences are psychologically unrealistic (Jeffrey 1983; Kyburg 1983; Kaplan 1996), that only indeterminate credences are warranted in many epistemic situations that feature insufficient or imprecise evidence (Levi 1974, 1980; Walley 1991; Joyce 2005, 2010; Sturgeon 2008), or that a group of agents is usually in an epistemic situation that can be aptly modeled by imprecise probabilities (Walley 1991; Seidenfeld et al. 1995)—are also applicable here. Our observation above is that in the context of subjective causal networks, there appears to be a good reason in addition to these standard arguments for accommodating imprecise probabilities.

[4] A closely related notion known as 'strong independence' applies to convex sets of probability functions. Two variables X and Y are *strongly independent* if the set of probability functions is the convex hull of a set that satisfies complete independence (Cozman 2012). As we intimated earlier, we have no intention to enforce the requirement of convexity for credal sets in this paper.

[5] In a way, this is similar to what Spirtes et al. (2000, pp. 42–43) suggested in their brief discussion of "Bayesian interpretations" of the formalism. However, in their suggestion the set of probability functions is explicitly indexed by a set of parameters that, following orthodox Bayesianism, are also treated as random variables.

The move toward indeterminacy is of course not entirely uncontroversial. Elga (2010), for example, argued that the credences of a perfectly rational agent should always be sharp, but how compelling the argument is has been forcefully disputed (Bradley and Steele 2014; Chandler 2014). Indeterminacy also complicates, and sometimes appears to stymie reasoning. For example, if an agent's doxastic state about a coin's bias is maximally uninformative, as represented by the entire interval [0, 1], then the standard Bayesian updating on the corresponding set of probability functions (by updating each member of the set) will always yield the same set, come what may. However, this apparent impossibility to learn from experience may be entirely appropriate; it is arguable that any informative learning in such a situation involves an unwarranted leap. Alternatively, for those who think learning should be possible even in such extreme cases, it is also possible to devise reasonable updating policies that enable it (Joyce 2010).[6] Since the focus of this paper is on the interpretation and foundations of imprecise probabilities, we refer readers who are interested in reasoning and inference to (Augustin et al. 2014).

To summarize, despite unresolved debates, it is at least worthwhile to explore causal networks with imprecise probabilities. Once imprecise probabilities are allowed, with an appropriate notion of (conditional) independence, the causal network in question is not a CBN any more, but a generalization of CBN, which we describe next.

2.3 Causal credal networks

In the artificial intelligence literature, there are generalizations of BNs that allow imprecise probabilities, known as credal networks (Cozman 2000; Cozman and Walley 2001; De Bock and de Cooman 2015; De Bock 2017). Among the several types of credal networks that have been studied, the one that is most straightforward and most relevant to addressing the previous issue can be defined as follows: A *credal network (CN)* over $\mathbf{V}$ is a pair $(\mathcal{G}, \mathbf{K})$, where $\mathcal{G}$ is a DAG and $\mathbf{K}$ is a non-empty set of joint probability distributions such that *every* member of $\mathbf{K}$ is Markov to $\mathcal{G}$.[7] An extremely simple example of a CN was alluded to previously, where $\mathcal{G}$ is a DAG over two

6 We thank an anonymous referee for stressing these points. The referee also raises the following question. What if the agent has a prior or higher-order probability distribution over the set of probability functions? Should the prior be ignored or should it be used to integrate out parameters to form a single, precise probability function? This is an important question to which we cannot do justice here, in part because we are yet to make up our (group) mind about it. However, it is fair to say that for the reason elaborated earlier, unless we give up the subjective Markov condition, we cannot simply collapse the hierarchical structure into a precise distribution over the substantive variables and treat that single distribution as all there is to the agent's doxastic state (though this single distribution may well be useful for other purposes). On the other hand, to ignore the prior seems obviously unreasonable. Presumably it is the whole hierarchical structure, consisting of the set and the distribution over the set, that is a proper representation of the credences in question, in which case a subjective Markov condition in some form may still be maintained (cf. Spirtes et al. 2000, pp. 42–43, as is mentioned in the previous note). A broadly similar representation was used in (Gärdenfors and Sahlin 1982), where each member of a set of probability functions is labeled with a degree of epistemic reliability. Our results in this paper do not cover credences with this additional complexity.

7 This corresponds to what is known as credal networks under complete independence. Two more commonly studied types are known as credal networks under strong independence, which is, roughly speaking, a convexized version of that under complete independence, and credal networks under epistemic irrelevance, which is importantly different (De Bock 2017).

variables X and Y without any arrow, and $\mathbf{K}$ is a set that contains two joint probability distributions over X and Y, each of which factorizes according to $\mathcal{G}$ (i.e., factorizes into a product of two marginals).[8] Formally, a CN (as we define it here) is equivalent to a set of BNs with the same graph, and a BN is clearly a special case of a CN (in which the set $\mathbf{K}$ is a singleton).

In the same fashion, we can define *causal credal networks (CCNs)* as a generalization of CBNs:

Definition 2 (*Causal credal networks*) A causal credal network over $\mathbf{V}$ is a pair $(\mathcal{G}, \mathbf{CK})$, where $\mathcal{G}$ is a DAG over $\mathbf{V}$, and $\mathbf{CK}$ is a non-empty set of collections of interventional probabilities, such that every collection in the set, $\{P^{do(\mathbf{S}=\mathbf{s})}\}$, satisfies the following intervention principle with $\mathcal{G}$: for every $\mathbf{S} \subseteq \mathbf{V}$ and every possible value configuration $\mathbf{s}$ for $\mathbf{S}$,

$$P^{do(\mathbf{S}=\mathbf{s})}(\mathbf{V}-\mathbf{S}) = \prod_{X \in \mathbf{V}-\mathbf{S}} P(X|\mathbf{Pa}_{\mathcal{G}}(X)),$$

(provided that the conditional probabilities on the right-hand side are all defined), where $P = P^{do(\emptyset)}$ is the observational distribution in the collection, and for each $P(X|\mathbf{Pa}_{\mathcal{G}}(X))$ on the right-hand side, every $Y \in \mathbf{S} \cap \mathbf{Pa}_{\mathcal{G}}(X)$ is assigned the value that is consistent with $\mathbf{S} = \mathbf{s}$.

Again, a CCN can be regarded as a set of CBNs with the same causal graph, and a CBN is a special case of a CCN (in which the set $\mathbf{CK}$ contains only one collection of interventional probability distributions). Propositions 1 and 2 are also easily generalizable (provided that the observational distribution in every collection in $\mathbf{CK}$ is strictly positive).

Proposition 3 *Let* $(\mathcal{G}, \mathbf{CK})$ *be a causal credal network over* $\mathbf{V}$. *For every* $X \neq Y \in \mathbf{V}$, *and* $\mathbf{Z} = \mathbf{V} - \{X, Y\}$, *if there exist* $\{P^{do(\mathbf{S}=\mathbf{s})}\} \in \mathbf{CK}$, *two distinct values,* $x_1 \neq x_2$, *for* X *and a value configuration* $\mathbf{z}$ *for* $\mathbf{Z}$, *such that*

$$P^{do(X=x_1 \wedge \mathbf{Z}=\mathbf{z})}(Y) \neq P^{do(X=x_2 \wedge \mathbf{Z}=\mathbf{z})}(Y),$$

then there is an arrow from X *to* Y *in* $\mathcal{G}$.

Proposition 4 *Let* $(\mathcal{G}, \mathbf{CK})$ *be a causal credal network over* $\mathbf{V}$, *and let* $\mathcal{G}_I$ *be the graph over* $\mathbf{V}$ *that contains exactly the arrows that satisfy the condition in Proposition 3. Then* $(\mathcal{G}_I, \mathbf{CK})$ *is also a causal credal network over* $\mathbf{V}$.

Propositions 3 and 4 motivate a generalization of the interventionist definition of causal structure to the context of CCN: X is directly causally relevant to Y just in case

[8] Notice that in this example, the marginal of X contains two distributions, and so does the marginal of Y. However, the set $\mathbf{K}$ contains only two *joint* distributions of X and Y, rather than all four joint distributions that can be generated from the two marginal sets. That is, in our conception of a credal network, we do not require that the credal set include every joint distribution that can be generated by picking one member from each local marginal or conditional, and taking their product.

the condition in Proposition 3 is met. We will therefore restrict our attention to those CCNs in which the graph accurately represents this notion of causal structure.

We hasten to note that unlike the case with precise probabilities, in the context of imprecise probabilities there appear to be more than one intuitively meaningful notion of direct causal relevance.[9] For example, in addition to the above definition, it may be natural to consider a stronger notion that can be defined by replacing the existential quantifier in the condition in Proposition 3 with a universal quantifier. Our observation here is that the weaker notion is, so to speak, picked out by the formalism of CCN, in the same way that the standard interventionist notion is picked out by the formalism of CBN. In other words, the arrows in a CCN should be interpreted as representing the causal notion defined by the condition in Proposition 3, even though it may not be the notion that most deserves the label of 'direct causal relevance'.

More importantly for present purposes, with this well motivated definition of direct causal relevance, the pair that constitutes a CCN is reducible to its second element. This point, as we argued previously, relieves the burden of providing an independent account of qualitative causal beliefs in a subjective interpretation of CCNs. To repeat, there may be other interesting notions of direct causal relevance in addition to the notion that we have just shown is naturally attached to arrows in a CCN. If we understand a causal structure as a set of qualitative statements of direct causal relevance/irrelevance, then there may be other interesting notions of causal structure in addition to the one that is represented in a CCN. Still, we contend that in any interesting sense of 'causal structure', what it is for an agent to believe a causal structure is ultimately a matter of her suppositional credences. The sense represented by arrows in a CCN is just a case in point.

Another limitation of our focus on the CCN framework is that a CCN is a set of CBNs of a special kind, a set whose members share the same causal graph. If a CCN is used to represent an agent's doxastic situation, the graph is interpreted as encoding the causal structure believed by the agent. But of course an agent need not believe a specific causal structure; it is commonplace to have uncertainty about the causal structure itself. To model such uncertainty along the line we have been considering, it is natural to consider a set of causal networks that do not necessarily share the same causal structure. In such a situation, it is a set of causal structures that corresponds to the agent's doxastic state. Such a set, we speculate, will be similarly reduced to suppositional credences.[10]

2.4 Indeterminate suppositional credences

Consider again the simple example described in Sect. 2.2. An agent believes that two causally unconnected coin flips nonetheless share the same chance of landing heads,

[9] This reflects a rather common theme in the study of imprecise probabilities. Another most salient example is that the notion of stochastic independence is fruitfully generalizable in more than one way in the context of imprecise probabilities, as we mentioned before. We thank an anonymous referee for making this observation.

[10] We thank an anonymous referee for pressing us on this point. Moreover, as the referee pointed out, it may be worth considering credal networks of an altogether different type, such as credal networks under epistemic irrelevance.

and that the chance is either 0.4 or 0.6. As explained previously, in this case the agent's epistemic state can hardly be properly modeled by a CBN, but seems to fit a CCN in a straightforward way. More specifically, it can be represented by a CCN, $(\mathcal{G}, \mathbf{CK})$, over two (binary) variables X and Y representing the outcomes of the two coin flips, respectively. $\mathcal{G}$ is the empty graph over X and Y, i.e., a graph with no arrow between them, and is taken to represent the causal structure that is believed by the agent. $\mathbf{CK}$ is taken to represent the agent's (indeterminate) suppositional credences, and formally contains two members, $\{P\}_i$ $(i = 1, 2)$, each of which is a collection of interventional probability distributions. $\{P\}_1$ contains the following distributions:

$$P_1^{do(\emptyset)}(X, Y) = P_1^{do(\emptyset)}(X)P_1^{do(\emptyset)}(Y);\;\; P_1^{do(\emptyset)}(X = 1) = P_1^{do(\emptyset)}(Y = 1) = 0.4;$$
$$P_1^{do(X=1)}(Y = 1) = 0.4;$$
$$P_1^{do(X=0)}(Y = 1) = 0.4;$$
$$P_1^{do(Y=1)}(X = 1) = 0.4;$$
$$P_1^{do(Y=0)}(X = 1) = 0.4.$$

And $\{P\}_2$ contains the following:

$$P_2^{do(\emptyset)}(X, Y) = P_2^{do(\emptyset)}(X)P_2^{do(\emptyset)}(Y);\;\; P_2^{do(\emptyset)}(X = 1) = P_2^{do(\emptyset)}(Y = 1) = 0.6;$$
$$P_2^{do(X=1)}(Y = 1) = 0.6;$$
$$P_2^{do(X=0)}(Y = 1) = 0.6;$$
$$P_2^{do(Y=1)}(X = 1) = 0.6;$$
$$P_2^{do(Y=0)}(X = 1) = 0.6.$$

Notice that $\mathbf{CK}$ here is *non-convex*. For the reason mentioned in Sect. 2.2, adding any nontrivial, convex mixture of $\{P\}_1$ and $\{P\}_2$ into $\mathbf{CK}$ will violate our definition of a CCN. Therefore, in general we need to allow $\mathbf{CK}$ to be non-convex; convexity will be the exception rather than the rule in this context.

To illustrate the earlier point about the subjective causal structure, the belief that there is no arrow from X to Y is grounded in the indeterminate suppositional credences concerning Y given the supposition that X is intervened to take value 1 and given the supposition that X is intervened to take value 0. In this case these credences are represented by a set of two vectors of distributions:

$$\{\langle P_1^{do(X=1)}(Y), P_1^{do(X=0)}(Y)\rangle, \langle P_2^{do(X=1)}(Y), P_2^{do(X=0)}(Y)\rangle\}.$$

In each vector in the set, the distributions are identical, and it is this feature of the agent's credences that in our view constitutes her belief that X is not (directly) causally relevant to Y. Similarly for her belief that there is no arrow from Y to X.

More generally, whether an agent believes that X is directly causally relevant to Y relative to $\mathbf{V}$ is determined by her credences concerning Y given each supposition of the form: $\mathbf{S} = \mathbf{V} - \{Y\}$ is intervened to take value configuration $\mathbf{s}$. Suppose there are

m possible value configurations for $\mathbf{S}$, the credences in question are represented by a *set of vectors of distributions*: $\{\langle P^1(Y), \ldots, P^m(Y)\rangle\}$, where $P^i(Y)$ is shorthand for $P^{do(\mathbf{S}=\mathbf{s}_i)}(Y)$.

The proposed representation is a natural choice for a rather common kind of situation, in which the agent in question is certain about the presence of a causal arrow but uncertain about the magnitude (or even the sign) of the causal effect. Each vector in the set determines a possible magnitude (according to any quantitative measure proposed in the literature), and the set of vectors yields a set of possible magnitudes. Similarly, it is a natural choice for representing the kind of situation in which a group of experts agree on the qualitative causal structure but disagree on the magnitude of effects.

For example, suppose, in the simple two-variable setting above, **CK** actually contains the following two members:

$$\begin{aligned}
P_a^{do(\emptyset)}(X=1) &= 0.5;\\
P_a^{do(\emptyset)}(Y=1|X=1) &= 0.6;\\
P_a^{do(\emptyset)}(Y=1|X=0) &= 0.4;\\
P_a^{do(X=1)}(Y=1) &= 0.6;\\
P_a^{do(X=0)}(Y=1) &= 0.4;\\
P_a^{do(Y=1)}(X=1) &= 0.5;\\
P_a^{do(Y=0)}(X=1) &= 0.5.
\end{aligned}$$

and

$$\begin{aligned}
P_b^{do(\emptyset)}(X=1) &= 0.5;\\
P_b^{do(\emptyset)}(Y=1|X=1) &= 0.4;\\
P_b^{do(\emptyset)}(Y=1|X=0) &= 0.6;\\
P_b^{do(X=1)}(Y=1) &= 0.4;\\
P_b^{do(X=0)}(Y=1) &= 0.6;\\
P_b^{do(Y=1)}(X=1) &= 0.5;\\
P_b^{do(Y=0)}(X=1) &= 0.5.
\end{aligned}$$

This would then be a case in which the agent believes (or two agents agree) that X has a causal influence on Y (and not vice versa), but is uncertain (or two agents disagree) about the effect: she believes that X is causally relevant to Y, but for all she knows, the relevance might be positive (i.e., an action or intervention to set $X = 1$ increases the chance of $Y = 1$), or negative (i.e., an action or intervention to set $X = 1$ decreases the chance of $Y = 1$). Again, the agent's belief that X is causally relevant to Y is in our view constituted by her suppositional credences represented by the set

$$\{\langle P_a^{do(X=1)}(Y), P_a^{do(X=0)}(Y)\rangle, \langle P_b^{do(X=1)}(Y), P_b^{do(X=0)}(Y)\rangle\}.$$

Likewise for her belief that Y is not causally relevant to X.

Another potential use of the proposed representation is in the context of an agency or decision-theoretic approach to causation (Price 1991; Menzies and Price 1993; Hitchcock 1996). As mentioned earlier, what we have been calling suppositional credences are akin to agent probabilities in that approach, and the ambition is to account for causation (as opposed to beliefs about causation) in terms of these agent probabilities. It will take us too far afield to evaluate the promise of this approach here, but we submit that for anyone who is sympathetic with this approach and at the same time is convinced by our argument above and other arguments in the literature for taking indeterminate credences seriously, a strong contender for representing *indeterminate* agent probabilities is such sets of vectors of probability functions as $\{\langle P^1(Y), \ldots, P^m(Y)\rangle\}$.

In the second part of the paper, we will focus on such indeterminate suppositional credences as represented by $\{\langle P^1(Y), \ldots, P^m(Y)\rangle\}$, and try to "capture" them with a generalized theory of coherent choice functions. It is natural to think of each $do(\mathbf{S} = \mathbf{s}_i)$ as a possible act and each possible value of Y as a possible state. With some prizes/consequences, we can construct decision situations on which the indeterminate suppositional credences will have bearings. Such a decision situation, however, will have to allow act-state dependence. In the above example with P_a and P_b, for instance, the natural decision problem on which $\{\langle P_a^{do(X=1)}(Y), P_a^{do(X=0)}(Y)\rangle, \langle P_b^{do(X=1)}(Y), P_b^{do(X=0)}(Y)\rangle\}$ bears will feature act-state dependence: states $Y = 1$ and $Y = 0$ are not independent of actions $do(X = 1)$ and $do(X = 0)$. In addition, the envisaged decision theory should allow credences to be non-convex sets of (vectors of) probabilities. To our knowledge, no such decision theory is yet available.

3 A generalization of Anscombe and Aumann's theory of horse lotteries

Seidenfeld, Schervish and Kadane (SSK)'s theory of coherent choice functions builds on the framework of 'horse lotteries' introduced by Anscombe and Aumann (AA, 1963). In this section, we present a generalization of AA's celebrated and elegant decision-theoretic "definition" of (precise) subjective probability. The main purpose is to provide a relatively simple illustration of the general strategy of our generalization and how it is supposed to work in the context of precise probability, before moving on to the more complex context of imprecise probability.

Let $\Omega = \{w_1, \ldots, w_n\}$ be a finite space of states and $\mathcal{R}$ be a denumerable set of prizes. Following AA and SSK, we assume that $\mathcal{R}$ contains a most desired prize B ('best') and a least desired prize W ('worst'), between which there is a strict preference. A *von Neumann–Morgenstern (vNM) lottery* is an (objective or extraneous) probability distribution over the prizes (von Neumann and Morgenstern 1947). Denote the set of vNM lotteries with $\mathcal{L}$. An Anscombe–Aumann *horse lottery* is a function from Ω to $\mathcal{L}$.

Given any vNM lotteries l_1 and l_2, for $0 \leq \alpha \leq 1$, we write $\alpha l_1 \oplus (1-\alpha) l_2$ to denote a convex combination of l_1 and l_2, which is itself a vNM lottery (i.e., a probability distribution over the prize set). Similarly, for any horse lotteries, which we will also call *options*, o_1 and o_2, and $0 \le \alpha \le 1$, we write $\alpha o_1 \oplus (1-\alpha)o_2$ to denote the horse lottery, o_3, resulting from state-wise convex combination of o_1 and o_2; that is, for every $w \in \Omega$, $o_3(w) = \alpha o_1(w) \oplus (1-\alpha)o_2(w)$.

Let $\mathcal{H}$ denote the set of horse lotteries. AA's main result is a characterization of what we may call coherent (binary) preference relations over $\mathcal{H}$. Let $\preceq$ denote a binary (weak) preference relation over $\mathcal{H}$ ('$o_1 \preceq o_2$' may be informally read as 'o_1 is less desirable than or as desirable as o_2'). The relation $\preceq$ is *coherent* just in case there exist a probability distribution P over Ω and a (state-independent) linear utility function u over $\mathcal{L}$ that together *represent* $\preceq$, in the sense that for every $o_1, o_2 \in \mathcal{H}$, $o_1 \preceq o_2$ if and only if $\sum_{w\in\Omega} P(w)u(o_1(w)) \le \sum_{w\in\Omega} P(w)u(o_2(w))$ (i.e., the (P, u)-expected utility of o_1 is less than or equal to the (P, u)-expected utility of o_2.)

AA's main result implies that every coherent preference relation in this sense is represented by a *unique* probability distribution over Ω, which they take to be the subjective probability of the agent who subscribes to the preference relation. As AA showed, coherent preference relations can be elegantly axiomatized. Below is a version of the axiomatization (Seidenfeld et al. 2010). A *constant* horse lottery assigns the same vNM lottery to each state in Ω. For convenience we introduce the following notation for a special kind of constant horse lottery. For every prize $r_i \in \mathcal{R}$, we use $\mathbf{R}_i$ to denote the constant horse lottery that assigns to each state in Ω the vNM lottery rewarding r_i for certain (i.e., the constant vNM lottery rewarding r_i). In particular, $\mathbf{W}$ denotes the constant horse lottery of this special kind corresponding to prize W and $\mathbf{B}$ denotes the constant horse lottery of this special kind corresponding to prize B. Also, let $\prec$ denote the asymmetric part of $\preceq$: $o_1 \prec o_2$ just in case $o_1 \preceq o_2$ but not vice versa. We first make explicit the assumption that there is a best (B) and a worst (W) in the prize set:

AA0: $\mathbf{W} \prec \mathbf{B}$, and for every constant horse lottery $\mathbf{R}$, $\mathbf{W} \preceq \mathbf{R} \preceq \mathbf{B}$.

Next is an axiom stating that the preference relation generates a weak order.

AA1: $\preceq$ is a weak order; that is, it is reflexive, transitive, and complete.

The next axiom is usually known as the axiom of independence:

AA2: For every o_1, o_2, o_3, and every $0 < \alpha \le 1$, $o_1 \prec o_2$ iff $\alpha o_1 \oplus (1-\alpha)o_3 \prec \alpha o_2 \oplus (1-\alpha)o_3$.

Then there is an Archimedean condition:

AA3: For every o_1, o_2, o_3, if $o_1 \prec o_2 \prec o_3$, then there exist $0 < \alpha, \beta < 1$ such that $\alpha o_1 \oplus (1-\alpha)o_3 \prec o_2 \prec \beta o_1 \oplus (1-\beta)o_3$.

Finally, an axiom is needed to ensure the existence of a representation with state-independent utilities. Let $\sim$ denote the symmetric part of $\preceq$: $o_1 \sim o_2$ just in case $o_1 \preceq o_2$ and $o_2 \preceq o_1$. A state $w \in \Omega$ is said to be *null* if for every $o_1, o_2 \in \mathcal{H}$ such that $o_1(w') = o_2(w')$ for all $w' \neq w$, $o_1 \sim o_2$. Otherwise w is called *non-null*.

AA4: Let w be a non-null state and $r_1, r_2 \in \mathcal{R}$ be any prizes. For every o_1, o_2 such that $o_1(w) = r_1$ (the vNM lottery that rewards r_1 for certain), $o_2(w) = r_2$ (the vNM lottery that rewards r_2 for certain), and $o_1(w') = o_2(w')$ for all $w' \neq w$, $o_1 \preceq o_2$ if and only if $\mathbf{R}_1 \preceq \mathbf{R}_2$.

In plain words, **AA4** states that the preference between any two prizes remains the same in all non-null states.

AA's main result implies that axioms **AA0–AA4** are necessary and sufficient for $\preceq$ to be coherent (under the assumption that W is a worst and B is a best prize in $\mathcal{R}$).

The promised generalization of AA's theory is fairly straightforward. Recall that the suppositional credences we are concerned with are represented by a set of vectors of probabilities. In the case of precise probabilities, the set is supposed to be a singleton, with a single vector $\langle P^1, \ldots, P^m \rangle$, where m is the number of alternative interventions or actions (to set the values of certain variables), and P^j is the probability distribution over the state space under the supposition of action j. P^j and P^k can be different for $j \neq k$, reflecting the possibility of act-state dependence.

To capture such a vector of probability distributions, we need to enrich the option space. For each horse lottery o, we write o^j $(1 \leq j \leq m)$ to denote the option of 'o given (supposed) intervention j'. Intuitively, to evaluate o^j is to evaluate o under the supposition that intervention j took place. The idea is very similar to the 'conditional' acts employed by Armendt (1986) and Joyce (1999), in order to handle act-state dependence.

Call these *indexed options* with index set $\{1, \ldots, m\}$. Let $\mathcal{H}^\dagger$ be the set of these indexed options. A binary preference relation $\preceq$ over $\mathcal{H}^\dagger$ is said to be *coherent* just in case there exist a vector of probability distributions over Ω, $\langle P^1, \ldots, P^m \rangle$, and a (state-independent and index-independent) linear utility function u over $\mathcal{L}$ that *represent* $\preceq$, in the sense that for every $o_1^j, o_2^k \in \mathcal{H}^\dagger$, $o_1^j \preceq o_2^k$ if and only if $\sum_{w \in \Omega} P^j(w)u(o_1^j(w)) \leq \sum_{w \in \Omega} P^k(w)u(o_2^k(w))$. Notice that options with different indices are evaluated according to (possibly) different probability distributions in the calculation of expected utility.

Given AA's uniqueness result, it trivially follows that every coherent preference relation over $\mathcal{H}^\dagger$ is represented by a *unique* vector of probability distributions, simply because each member of the vector can be uniquely recovered by AA's result from the restriction of the relation to options with the same index. Therefore, one possible way to provide a decision-theoretic basis for determinate suppositional credences is via coherent preference relations over indexed horse lotteries.

Moreover, coherent preference relations over $\mathcal{H}^\dagger$ can be axiomatized by a generalization of **AA0–AA4**, which we will label as **AA-I(ndexed)** axioms. A state $w \in \Omega$ is said to be *null with respect to index* j if for every $o_1^j, o_2^j \in \mathcal{H}^\dagger$ such that $o_1^j(w') = o_2^j(w')$ for all $w' \neq w$, $o_1^j \sim o_2^j$. Otherwise w is *non-null with respect to* j. Below are the **AA-I** axioms:

AA-I0: For every index j, $\mathbf{W}^j \prec \mathbf{B}^j$, and for every constant horse lottery $\mathbf{R}$, $\mathbf{W}^j \preceq \mathbf{R}^j \preceq \mathbf{B}^j$.

AA-I1: $\preceq$ is a weak order; that is, it is reflexive, transitive, and complete.

AA-I2: (a) For every o_1, o_2, o_3, every $0 < \alpha \leq 1$, and every index j, $o_1^j \prec o_2^j$ iff $\alpha o_1^j \oplus (1-\alpha)o_3^j \prec \alpha o_2^j \oplus (1-\alpha)o_3^j$.
(b) If o_1, o_2, o_3 are constant horse lotteries, then for every $0 < \alpha \leq 1$, and indices j, k, $o_1^j \prec o_2^k$ iff $\alpha o_1^j \oplus (1-\alpha)o_3^j \prec \alpha o_2^k \oplus (1-\alpha)o_3^k$.

AA-I3: For every o_1^j, o_2^k, o_3^j, if $o_1^j \prec o_2^k \prec o_3^j$, then there exist $0 < \alpha, \beta < 1$ such that $\alpha o_1^j \oplus (1-\alpha) o_3^j \prec o_2^k \prec \beta o_1^j \oplus (1-\beta) o_3^j$.

AA-I4: (a) Let w be a non-null state with respect to an index j, and $r_1, r_2 \in \mathcal{R}$ be any prizes. For every o_1^j, o_2^j such that $o_1^j(w) = r_1$, $o_2^j(w) = r_2$, and $o_1^j(w') = o_2^j(w')$ for all $w' \neq w$, $o_1^j \preceq o_2^j$ if and only if $\mathbf{R}_1^j \preceq \mathbf{R}_2^j$.
(b) For all j, k, $\mathbf{W}^j \sim \mathbf{W}^k$ and $\mathbf{B}^j \sim \mathbf{B}^k$.

Most axioms here are straightforward adaptations of the **AA** axioms to the space $\mathcal{H}^\dagger$. Note that axiom **AA-I2**(a) is a version of axiom **AA**2 that is restricted to options with the same index. An unrestricted formulation of the axiom of independence across different indices is in general not true of coherent preference relations, which is akin to the fact that the standard dominance principle fails in the presence of act-state dependence (Jeffrey 1965; Joyce 1999). However, a cross-index formulation that is restricted to constant horse lotteries remains sound and is presented above as **AA-I**2(b). There is also an extra clause in the last axiom: **AA-I**4(b) simply states that the desirability of the best prize and that of the worst prize are invariant across indices. Axioms **AA-I**2(b), **AA-I**3, and **AA-I**4(b) together ensure that there is a common (state-independent) linear utility (unique up to positive affine transformations) for vNM lotteries across different indices. A simple adaptation of AA's argument will then show that axioms **AA-I**0–**AA-I**4 are necessary and sufficient for a preference relation over $\mathcal{H}^\dagger$ to be coherent.

4 A generalization of (a simplified version of) SSK's theory of coherent choice functions

It is now time to consider an arbitrary set of vectors of probability distributions over Ω, which is our model for indeterminate suppositional credences that are related to subjective CCNs and to causal beliefs in general. Since we are not confined to convex sets of probabilities, we need to go beyond binary preference relations, for pairwise comparisons alone do not suffice to distinguish a set of probabilities from its convex hull (Seidenfeld et al. 1995). We will thus build on SSK's theory of coherent choice functions, which contains a sufficiently rich structure to distinguish arbitrary sets of probability distributions.[11] In this section we first outline (a simplified version of) SSK's theory, and then generalize it in the same fashion as we did for AA's theory. Given the set of horse lotteries $\mathcal{H}$ (on Ω and $\mathcal{R}$), a choice function C over $\mathcal{H}$ maps each closed subset of $\mathcal{H}$, O, to a subset of O, i.e., $C(O) \subseteq O$.[12] Intuitively, $C(O)$ is the set of options in O that are judged to be *acceptable* given the options in O. In SSK's theory, both probabilities and utilities are allowed to be indeterminate, and their target representation for a coherent choice function is in terms of a set of probability-utility pairs. For our present purpose, however, we are concerned only with indeterminacy in probability, and will not worry about indeterminacy in utility. Therefore, we are

[11] For a wealth of recent work on choice functions, see especially (Van Camp 2018; Van Camp et al. 2018; De Bock and de Cooman 2019).

[12] Here closure of a set of horse lotteries is defined with respect to the topology of state-wise convergence of distributions over the prize set (i.e., vNM lotteries).

interested in coherent choice functions relative to a common, state-independent utility function. In SSK's framework, these are choice functions that can be represented by a set $\mathbf{S} = \mathbf{Q} \times \{u\}$, where $\mathbf{Q}$ is a set of probability distributions over Ω, and u is a (state-independent) linear utility function over vNM lotteries. Here is the notion of coherence we need:

Definition 3 (*Coherent choice function*) A choice function C over $\mathcal{H}$ is coherent if there exist a non-empty set of probability distributions $\mathbf{Q}$ (over Ω) and a (state-independent) linear utility function u over $\mathcal{L}$ (the set of vNM lotteries) that *represent* C, in the sense that for every closed set of options $O \subseteq \mathcal{H}$ and $o \in O$, $o \in C(O)$ if and only if there exists $P \in \mathbf{Q}$, and for every $h \in O$,

$$\sum_{w\in\Omega} P(w)u(o(w)) \geq \sum_{w\in\Omega} P(w)u(h(w)).$$

In other words, for a coherent choice function, the acceptable options in any (closed) option set contain all and only those that maximize subjective expected utility relative to a utility function and for *some* probability distribution in a set. Note that a consequence of Definition 3 is that given a coherence choice function C, $C(O) \neq \emptyset$ for every *closed* set of options O.

An important result proved by SSK is that a coherent choice function according to Definition 3 is represented by a *unique* set of probability distributions over Ω. Thus, if we fix a linear utility function for the vNM lotteries, then there is a one-one correspondence between non-empty sets of probability distributions and coherent choice functions.

Following the strategy illustrated in the previous section, we now consider the space of indexed options $\mathcal{H}^\dagger$ (with index set $\{1, \ldots, m\}$). Consider choice functions over $\mathcal{H}^\dagger$. The notion of coherence is generalized as follows:

Definition 4 (*Coherent choice function over indexed options*) A choice function C over $\mathcal{H}^\dagger$ (with index set $\{1, ..., \mathrm{m}\}$) is coherent if there exist a non-empty set of vectors of probability distributions $\mathbf{Q}$ (over Ω) and a (state-independent and index-independent) linear utility function u over $\mathcal{L}$ that represent C, in the sense that for every closed set of options $O \subseteq \mathcal{H}^\dagger$ and $o^k \in O$, $o^k \in C(O)$ iff there exists $\langle P^1, ..., P^m \rangle \in \mathbf{Q}$, and for every $h^j \in O$,

$$\sum_{w\in\Omega} P^k(w)u(o^k(w)) \geq \sum_{w\in\Omega} P^j(w)u(h^j(w)).$$

Again, the main distinction in this generalization is that options with different indices are evaluated according to (possibly) different probability distributions. SSK's uniqueness result can be extended to this more general setting:

Proposition 5 *Every coherent choice function over indexed options is represented by a unique set of vectors of probability distributions.*

A proof is given in "Appendix". A contrast with the case of precise probability is worth noting. As we saw in the previous section, in the context of precise probability the uniqueness result on indexed horse lotteries is a trivial consequence of the uniqueness result on horse lotteries, because once the probability for each index is uniquely picked out, the vector is uniquely picked out. By contrast, although Proposition 5 is not hard to establish, it is not simply a trivial consequence of SSK's uniqueness result. It is true that SSK's uniqueness result immediately implies that given a coherence choice function over indexed options, for every index j, the restriction of the function to options with index j will pick out a unique set of probability distributions. However, putting these sets of probability distributions together yields a vector of sets of probabilities rather than a set of vectors of probabilities, the latter of which is what we aim to capture and is in general more fine-grained than the former.

Proposition 5 suggests that one possible way to provide a decision-theoretic basis for indeterminate suppositional credences is via coherent choice functions over indexed horse lotteries.

How to axiomatize coherent choice functions relative to a common, state-independent (and index-independent) utility remains an open problem. SSK's axiomatization does not ensure the existence of a representation with state-independent utilities, but employ instead a technical device called *almost state-independent utilities*. We now describe a set of axioms that adapt SSK's to the framework of indexed options, which we will refer to as indexed choice (**IC**) axioms. These axioms provide necessary conditions for coherent choice functions over indexed options.

Given a choice function C over $\mathcal{H}^\dagger$, define the corresponding rejection function: $R(O) = O - C(O)$. It is more convenient to state many axioms in terms of the rejection function. Again, we first make explicit the background assumption about prize W and prize B by restating **AA-I0** in terms of choice functions.

IC0: For every index j, $R(\{\mathbf{W}^j, \mathbf{B}^j\}) = \{\mathbf{W}^j\}$, and for every constant horse lottery $\mathbf{R}$, $\mathbf{R}^j \notin R(\{\mathbf{W}^j, \mathbf{R}^j\})$, and $\mathbf{B}^j \notin R(\{\mathbf{R}^j, \mathbf{B}^j\})$.

The next axiom can be regarded as a weakening of axiom **AA-I1**, the one that requires the preference relation to be a weak order. The first clause, **IC1**(a), is Sen's (1977) property α, and the second clause, **IC1**(b), is similar to Aizerman's (1985) condition.

IC1: Suppose O_1, O_2, O_3 are closed subsets of $\mathcal{H}^\dagger$. Then

(a) If $O_2 \subseteq R(O_1)$ and $O_1 \subseteq O_3$, then $O_2 \subseteq R(O_3)$.
(b) If $O_2 \subseteq R(O_1)$ and $O_3 \subseteq O_2$, then $O_2 - O_3 \subseteq R(\overline{O_1 - O_3})$, where $\overline{O_1 - O_3}$ denotes the closure of $O_1 - O_3$.

IC1 is related to ordering in the following sense. Given a rejection function R, define a binary relation over sets of options as follows: $O_1 \prec_R O_2$ just in case $O_1 \subseteq R(O_1 \cup O_2)$. Kadane et al. (2004) showed that if R satisfies **IC1**, then the relation $\prec_R$ is a strict partial order over sets of options. The next two axioms will be formulated in terms of $\prec_R$.

Let O^j denote an option set that contains only options with index j. Given an option set O^j and an option o^j, let $\alpha O^j \oplus (1-\alpha)o^j$ denote the set of element-wise convex mixtures, $\{\alpha h^j \oplus (1-\alpha)o^j \mid h^j \in O^j\}$. The next axiom is a generalization of the axiom of independence **AA-I2**.

IC2: (a) For $0 < \alpha \leq 1$, $O_1^j \prec_R O_2^j$ iff $\alpha O_1^j \oplus (1-\alpha)o^j \prec_R \alpha O_2^j \oplus (1-\alpha)o^j$.
(b) If O_1^j and O_2^k are sets of *constant* horse lotteries (with the respective indices) and o is a constant horse lottery, then for $0 < \alpha \leq 1$, $O_1^j \prec_R O_2^k$ iff $\alpha O_1^j \oplus (1-\alpha)o^j \prec_R \alpha O_2^k \oplus (1-\alpha)o^k$.

Like **AA-I2**, clause (a) of **IC2** restricts the application of the axiom of independence to options with the same index, and clause (b) adds that for constant horse lotteries, the axiom can apply across indices.

An analogue to **AA-I3**, the Archimedean axiom, is the following postulate of continuity:

IC3: Suppose O_{1k} and O_{2k} ($k = 1, ...$) converge element-wise to O_1 and O_2, respectively, and O is any (closed) option set. Then

(a) If for each k, $O_{1k} \prec_R O_{2k}$ and $O_2 \prec_R O$, then $O_1 \prec_R O$.
(b) If for each k, $O_{1k} \prec_R O_{2k}$ and $O \prec_R O_1$, then $O \prec_R O_2$.

To state an analogue to **AA-I4**, we need the following notion. Consider horse lotteries of the following form: $o_i(w_j) = \beta_{ij}B + (1-\beta_{ij})W$. We say that o_2 *weakly dominates* o_1 just in case $\beta_{2j} \geq \beta_{1j}$ for all j. The following axiom states a sort of state-neutrality regarding the preference between the prizes.

IC4: Suppose o_2^j weakly dominates o_1^j. Then for any $o_1^j \neq o^k \neq o_2^j$, if $o^k \in R(\{o_1^j\} \cup O)$, then $o^k \in R(\{o_2^j\} \cup O - \{o_1^j\})$.[13]

Intuitively, **IC4** states that replacing an option with a weakly dominating one does not promote any other unacceptable option to an acceptable one.

There is another axiom about mixtures in SSK's theory. Let $\mathbf{H}(O)$ denote the closed convex hull of a set of non-indexed options O. The axiom states that for every $o \in O$, if $o \in R(\mathbf{H}(O))$, then $o \in R(O)$. In other words, adding mixtures to an option set does not turn an originally acceptable option into an unacceptable one.

Note that the space of indexed options does not contain any mixture of options with different indices. In **IC2** (and **AA-I2**), for example, we only consider mixtures of options with the same index. Therefore, to adapt the above axiom to the setting of indexed options, we need to slightly adjust the meaning of $\mathbf{H}(O)$. In general, any set of indexed options O can be partitioned into a number of option sets such that (1) each member of the partition contains options with the same index, and (2) different members contain options with difference indices. Let $\mathbf{H}^*(O)$ be the union of the closed convex hulls of the members of the partition. (Obviously, if O contains only options with the same index, then $\mathbf{H}^*(O) = \mathbf{H}(O)$.) We then have

IC5: For every option set O and $o \in O$, if $o \in R(\mathbf{H}^*(O))$, then $o \in R(O)$.

Axioms **IC1**–**IC5** are adaptations of SSK's axioms. It is a routine exercise to check that they remain sound for coherent choice functions over indexed options. They are,

[13] In Seidenfeld et al. (2010), the non-indexed version of this axiom is formulated as two clauses:

(a): If $o_2 \in O$ and $o \in R(\{o_1\} \cup O)$, then $o \in R(O)$, and
(b): If $o_1 \in O$ and $o \in R(O)$, then $o \in R(\{o_2\} \cup O - \{o_1\})$.

It is not hard to see that the two clauses are equivalent to the simpler formulation: for any $o_1 \neq o \neq o_2$, if $o \in R(\{o_1\} \cup O)$, then $o \in R(\{o_2\} \cup O - \{o_1\})$. **IC4** is an adaptation of this simpler formulation.

however, insufficient to ensure the existence of a representation with a common, state-independent and index-independent utility, and hence insufficient for coherence in the sense of Definition 4. This is a gap to be filled in future work.[14]

5 Conclusion

Motivated by the aim to make sense of a subjective interpretation of a popular framework for causal modeling in artificial intelligence, we have, in the first part of the paper, highlighted the potential significance of indeterminate suppositional credences and proposed a formal representation of such credences in terms of sets of vectors of probability functions. Then, in the second part of the paper, we explored the possibility of providing a decision-theoretic foundation for such indeterminate suppositional credences. We outlined a generalization of the theory of coherent choice functions over horse lotteries, and showed that the generalized theory has enough structures to pick out indeterminate suppositional credences.

What we have done here is only a first step toward a decision-theoretic account of subjective causal (Bayesian or credal) networks. We have not tried to link the intervention principle itself to choice patterns, the feasibility of which remains to be explored. What we have tried to show is a way to link an arrow, or causal relevance in general, in a subjective causal network to choice patterns. We have taken a subjective causal network to represent beliefs about causation; for us, to make sense of causal relevance in a subjective causal network is to make sense of what it is for an agent to believe that a certain variable is causally relevant to another variable. But again, our framework may be easily adapted for the purpose of developing an agency or decision-theoretic account of causation with indeterminate credences.

The decision theory we are beginning to develop is undoubtedly too complex to be uncontroversial, but we do not yet see any simpler alternative that can potentially provide a decision-theoretic foundation for the kind of credences we highlighted in this paper. In any case, a decision theory that allows both imprecise probabilities and act-state dependence is probably interesting in its own right.

Acknowledgements This research was supported by the Research Grants Council of Hong Kong under the General Research Fund LU13600715, and by a Faculty Research Grant from Lingnan University. The research of the third author was supported by the Humanity and Social Science Youth Foundation of Ministry of Education of China (No. 16YJC72040001) and the MOE Project of Key Research Institute of Humanities and Social Sciences at Universities (No .17JJD720008).

[14] We suspect that we will need to add at least:

IC6: For every pair of constant horse lotteries o_1^j and o_2^k, if $C(\{o_1^j, o_2^k\}) = \{o_1^j, o_2^k\}$, then for any closed option set $O \supseteq \{o_1^j, o_2^k\}$, $o_1^j \in C(O)$ if and only if $o_2^k \in C(O)$.

This is Sen's (1977) property β restricted to constant horse lotteries. Since our notion of a coherent choice function is relative to a common, state-independent and index-independent utility, this axiom is obviously sound. Together with **IC**1(a), this axiom ensures an index-independent weak ordering of the prizes, which we believe is needed for the desired representation theorem.

Appendix

In this appendix, we prove Propositions 1–5. A proof of Proposition 1 and Proposition 2 was given in Zhang and Spirtes (2011). We include a proof here for the sake of Propositions 3 and 4.

Proof of Proposition 1 Suppose for the sake of contradiction that there exist $X, Y \in \mathbf{V}$ such that there is no arrow from X to Y in $\mathcal{G}$ (i.e., $X \notin \mathbf{Pa}_{\mathcal{G}}(Y)$), but there exist two distinct possible values, $x_1 \neq x_2$, for X and a value configuration $\mathbf{z}$ for $\mathbf{Z} = \mathbf{V} - \{X, Y\}$ such that

$$P^{do(X=x_1 \wedge \mathbf{Z}=\mathbf{z})}(Y) \neq P^{do(X=x_2 \wedge \mathbf{Z}=\mathbf{z})}(Y).$$

Since $X \notin \mathbf{Pa}_{\mathcal{G}}(Y)$, $\mathbf{Pa}_{\mathcal{G}}(Y) \subseteq \mathbf{Z}$. Then, since according to the intervention principle that defines the CBN, we have

$$P^{do(X=x_1 \wedge \mathbf{Z}=\mathbf{z})}(Y) = P^{do(\emptyset)}(Y|\mathbf{Pa}_{\mathcal{G}}(Y) = \mathbf{s}) = P^{do(X=x_2 \wedge \mathbf{Z}=\mathbf{z})}(Y),$$

where $\mathbf{s}$ is the value configuration of $\mathbf{Pa}_{\mathcal{G}}(Y)$ that is consistent with $\mathbf{Z} = \mathbf{z}$. (Recall that $P^{do(\emptyset)}$ is assumed to be positive, so $P^{do(\emptyset)}(Y|\mathbf{Pa}_{\mathcal{G}}(Y) = \mathbf{s})$ is defined.) Contradiction. □

Proof of Proposition 2[15]

Proof Since $(\mathcal{G}, \{P^{do(\mathbf{S}=\mathbf{s})}\})$ is a CBN over $\mathbf{V}$, by Proposition 1, it contains $\mathcal{G}_I$ as a subgraph. If $\mathcal{G} = \mathcal{G}_I$, the desired conclusion follows. Otherwise, $\mathcal{G}$ contains arrows that do not satisfy the condition in Proposition 1. Since $\mathcal{G}$ is finite, the number of such arrows is finite. Let these arrows be ordered as $E_1, ..., E_m$, and let $\mathcal{G}_j$ $(1 \leq j \leq m)$ denote the graph resulting from taking away the set of arrows $\{E_1, ..., E_j\}$ from $\mathcal{G}$. Obviously $\mathcal{G}_m = \mathcal{G}_I$. Let $\mathcal{G}_0 = \mathcal{G}$. We show by induction that for all $0 \leq j \leq m$, $(\mathcal{G}_j, \{P^{do(\mathbf{S}=\mathbf{s})}\})$ is a CBN over $\mathbf{V}$.

The base case is trivial. Suppose $(\mathcal{G}_k, \{P^{do(\mathbf{S}=\mathbf{s})}\})$ is a CBN over $\mathbf{V}$, we now argue that $(\mathcal{G}_{k+1}, \{P^{do(\mathbf{S}=\mathbf{s})}\})$ is also a CBN over $\mathbf{V}$. Note that the only difference between $\mathcal{G}_k$ and $\mathcal{G}_{k+1}$ is that the former, but not the latter, contains the arrow E_{k+1}. Let $X \rightarrow Y$ denote the arrow E_{k+1}. To show that $(\mathcal{G}_{k+1}, \{P^{do(\mathbf{S}=\mathbf{s})}\})$ is a CBN, it suffices to show that for every $V \in \mathbf{V}$, $P^{do(\emptyset)}(V|\mathbf{Pa}_{\mathcal{G}_{k+1}}(V)) = P^{do(\emptyset)}(V|\mathbf{Pa}_{\mathcal{G}_k}(V))$, because $(\mathcal{G}_k, \{P^{do(\mathbf{S}=\mathbf{s})}\})$ is a CBN by the inductive hypothesis. This is obviously the case for every $V \neq Y$, which has the exact same parents in both $\mathcal{G}_{k+1}$ and $\mathcal{G}_k$. Regarding Y, $\mathbf{Pa}_{\mathcal{G}_k}(Y) = \mathbf{Pa}_{\mathcal{G}_{k+1}}(Y) \cup \{X\}$. Since the arrow $X \rightarrow Y$ does not satisfy the condition in Proposition 1, it follows that for every value configuration $\mathbf{s}$ for $\mathbf{S} = \mathbf{Pa}_{\mathcal{G}_{k+1}}(Y)$, value configuration $\mathbf{t}$ for $\mathbf{T} = \mathbf{V} - \{X, Y\} - \mathbf{Pa}_{\mathcal{G}_{k+1}}(Y)$, and distinct values $x_1 \neq x_2$ for X, we have

$$P^{do(X=x_1 \wedge \mathbf{S}=\mathbf{s} \wedge \mathbf{T}=\mathbf{t})}(Y) = P^{do(X=x_2 \wedge \mathbf{S}=\mathbf{s} \wedge \mathbf{T}=\mathbf{t})}(Y).$$

[15] We thank an anonymous referee for suggesting this direct proof rather than an indirect one as it was originally presented.

On the other hand, since $(\mathcal{G}_k, \{P^{do(\mathbf{S}=\mathbf{s})}\})$ is a CBN, for both $i = 1, 2$ we have

$$P^{do(X=x_i \wedge \mathbf{S}=\mathbf{s} \wedge \mathbf{T}=\mathbf{t})}(Y) = P^{do(\emptyset)}(Y|X = x_i \wedge \mathbf{S} = \mathbf{s}).$$

because $\mathbf{Pa}_{\mathcal{G}_k}(Y) = \mathbf{S} \cup \{X\}$. Hence, for any distinct values x_1, x_2 for X and any value configuration $\mathbf{s}$ for $\mathbf{Pa}_{\mathcal{G}_{k+1}}(Y)$,

$$P^{do(\emptyset)}(Y|X = x_1 \wedge \mathbf{Pa}_{\mathcal{G}_{k+1}}(Y) = \mathbf{s}) = P^{do(\emptyset)}(Y|X = x_2 \wedge \mathbf{Pa}_{\mathcal{G}_{k+1}}(Y) = \mathbf{s}).$$

Thus, $P^{do(\emptyset)}(Y|\mathbf{Pa}_{\mathcal{G}_k}(Y)) = P^{do(\emptyset)}(Y|\mathbf{Pa}_{\mathcal{G}_{k+1}}(Y) \cup \{X\}) = P^{do(\emptyset)}(Y|\mathbf{Pa}_{\mathcal{G}_{k+1}}(Y))$. Therefore, for every $V \in \mathbf{V}$, $P^{do(\emptyset)}(V|\mathbf{Pa}_{\mathcal{G}_{k+1}}(V)) = P^{do(\emptyset)}(V|\mathbf{Pa}_{\mathcal{G}_k}(V))$, which implies that $(\mathcal{G}_{k+1}, \{P^{do(\mathbf{S}=\mathbf{s})}\})$ is also a CBN.

Therefore, $(\mathcal{G}_j, \{P^{do(\mathbf{S}=\mathbf{s})}\})$ is a CBN over $\mathbf{V}$ for all $0 \leq j \leq m$. The desired conclusion follows since $\mathcal{G}_m = \mathcal{G}_I$. □

Proof of Proposition 3 By the definition of a CCN (Definition 2), for every $\{P^{do(\mathbf{S}=\mathbf{s})}\} \in \mathbf{CK}$, $(\mathcal{G}, \{P^{do(\mathbf{S}=\mathbf{s})}\})$ is a CBN. Then the argument for Proposition 1 can be used here to establish the desired conclusion.

Proof of Proposition 4 Again, for every $\{P^{do(\mathbf{S}=\mathbf{s})}\} \in \mathbf{CK}$, $(\mathcal{G}, \{P^{do(\mathbf{S}=\mathbf{s})}\})$ is a CBN. Therefore, if an arrow in $\mathcal{G}$ does not satisfy the condition in Proposition 3, then the argument for Proposition 2 can be applied for every $\{P^{do(\mathbf{S}=\mathbf{s})}\} \in \mathbf{CK}$, establishing that the DAG resulting from taking away the arrow from $\mathcal{G}$ still constitutes a CBN with every $\{P^{do(\mathbf{S}=\mathbf{s})}\} \in \mathbf{CK}$, and so still constitutes a CCN with $\mathbf{CK}$. □

Proof of Proposition 5 The proof is a generalization of a simplified version of SSK's argument for their uniqueness result (Seidenfeld et al. 2010, pp. 161–162). Consider an arbitrary vector $\langle P^1, \ldots, P^m \rangle$ of distributions over Ω. Each P^j can itself be written as a probability vector with n coordinates. Let p_* be the smallest positive coordinate in the m probability vectors, and let $\underline{p} = \frac{1}{2}p_*$.[16] Let a denote the vNM lottery $\underline{p}B \oplus (1-\underline{p})W$. For each $k = 1, \ldots, n$ and $j = 1, \ldots, m$, define option o_k^j (i.e., option o_k with index j) as follows, where p_k^j denotes the kth coordinate of the probability vector P^j, namely, $P^j(w_k)$—the probability mass P^j assigns to state w_k:

$$o_k^j(\omega_i) = \begin{cases} B & \text{if } i = k \text{ and } p_k^j = 0, \\ a & \text{if } i \neq k \text{ and } p_k^j = 0, \\ \frac{\underline{p}}{p_k^j} B \oplus \left(1 - \frac{\underline{p}}{p_k^j}\right) W & \text{if } i = k \text{ and } p_k^j > 0, \\ W & \text{if } i \neq k \text{ and } p_k^j > 0. \end{cases}$$

Let $O_{\langle P^1, \ldots, P^m \rangle} = \{o_k^j \mid k = 1, \ldots, n;\ j = 1, \ldots, m\} \cup \{\mathbf{a}^1\}$, where $\mathbf{a}^1$ denotes the constant horse lottery (with index 1) that assigns to each state the vNM lottery a.

16 In SSK's argument, $\underline{p}$ is taken to be equal to p_*. The change here guarantees that $\underline{p} < 1$ and so avoids the need to consider the case of $\underline{p} = 1$ later in the argument.

Suppose C is a coherent choice function over $\mathcal{H}^\dagger$ that is represented by a set of vectors of probability distributions $\mathbf{Q}$ over Ω and a (state-independent and index-independent) utility function u over the set of vNM lotteries. Without loss of generality, suppose $u(B) = 1$ and $u(W) = 0$ (recall that a strict preference of B to W is assumed), so that $u(a) = \underline{p}$. We now show that $\langle P^1, \ldots, P^m \rangle \in \mathbf{Q}$ if and only if $\mathbf{a}^1 \in C(O_{\langle P^1,\ldots,P^m \rangle})$.

(**Only if**) Suppose $\langle P^1, \ldots, P^m \rangle \in \mathbf{Q}$. With respect to $\langle P^1, \ldots, P^m \rangle$ and u, the expected utility of $\{\mathbf{a}^1\}$ is obviously $u(a) = \underline{p}$. Consider now any $o_k^j \in O_{\langle P^1,\ldots,P^m \rangle}$, its expected utility with respect to $\langle P^1, \ldots, P^m \rangle$ and u is given by

$$\sum_{i=1}^{n} P^j(w_i)u(O_k^j(w_i)) = \begin{cases} u(a) = \underline{p} & \text{if } p_k^j = 0, \\ p_k^j u(O_k^j(w_k)) + (1 - p_k^j)u(W) = \underline{p} & \text{if } p_k^j > 0. \end{cases}$$

Thus, every option in $O_{\langle P^1,\ldots,P^m \rangle}$ has the same expected utility with respect to $\langle P^1, \ldots, P^m \rangle$ and u. Since C is represented by $\mathbf{Q}$ and u, and $\langle P^1, \ldots, P^m \rangle \in \mathbf{Q}$, $C(O_{\langle P^1,\ldots,P^m \rangle}) = O_{\langle P^1,\ldots,P^m \rangle}$, and so $\mathbf{a}^1 \in C(O_{\langle P^1,\ldots,P^m \rangle})$.

(**If**) Suppose $\mathbf{a}^1 \in C(O_{\langle P^1,\ldots,P^m \rangle})$. Consider any vector $\langle Q^1, \ldots, Q^m \rangle$ that is different from $\langle P^1, \ldots, P^m \rangle$. Then there exists $1 \le j \le m$ such that $P^j \ne Q^j$, which implies that there exists $1 \le k \le n$ such that $P^j(w_k) = p_k^j < q_k^j = Q^j(w_k)$. Note also that by definition, $\underline{p} < 1$. Then the expected utility of option o_k^j with respect to $\langle Q^1, \ldots, Q^m \rangle$ and u is given by

$$\sum_{i=1}^{n} Q^j(w_i)u(O_k^j(w_i)) = \begin{cases} q_k^j u(B) + (1 - q_k^j)u(a) > \underline{p} & \text{if } p_k^j = 0, \\ q_k^j u(O_k^j(w_k)) + (1 - q_k^j)u(W) = \frac{q_k^j}{p_k^j}\underline{p} > \underline{p} & \text{if } p_k^j > 0. \end{cases}$$

whereas the expected utility of $\mathbf{a}^1$ remains $\underline{p}$ with respect to $\langle Q^1, \ldots, Q^m \rangle$ and u. Therefore, for every $\langle Q^1, \ldots, Q^m \rangle \ne \langle P^1, \ldots, P^m \rangle$, $\mathbf{a}^1$ is not acceptable in $O_{\langle P^1,\ldots,P^m \rangle}$ with respect to $\langle Q^1, \ldots, Q^m \rangle$ and u. But by supposition, $\{\mathbf{a}^1\} \in C(O_{\langle P^1,\ldots,P^m \rangle})$, which means that $\{\mathbf{a}^1\}$ is acceptable with respect to $\mathbf{Q}$ and u. It follows that $\langle P^1, \ldots, P^m \rangle \in \mathbf{Q}$.

What we have shown is that whether an arbitrary vector of probability distributions is in the set that represents a given coherent choice function over $\mathcal{H}^\dagger$ is fully determined by the choice function. Therefore, the set of vectors of probability distributions that represents a given coherence choice function is unique. □

References

Aizerman, M. A. (1985). New problems in general choice theory. *Social Choice and Welfare*, *2*, 235–282.

Anscombe, F. J., & Aumann, R. J. (1963). A definition of subjective probability. *The Annals of Mathematical Statistics*, *34*, 199–205.

Armendt, B. (1986). A foundation for causal decision theory. *Topoi*, *5*, 3–19.

Augustin, T., Coolen, F. P. A., de Cooman, G., & Troffaes, M. C. M. (2014). *Introduction to imprecise probabilities*. Hoboken: Wiley.

Bradley, S., & Steele, K. (2014). Should subjective probabilities be sharp? *Episteme*, *11*, 277–289.

Chandler, J. (2014). Subjective probabilities need not be sharp. *Erkenntnis*, *79*(6), 1273–1286.
Couso, I., Moral, S., & Walley, P. (2000). A survey of concepts of independence for imprecise probabilities. *Risk, Decision and Policy*, *5*, 165–181.
Cozman, F. (2000). Credal networks. *Artificial Intelligence*, *120*, 199–233.
Cozman, F. (2012). Sets of probability distributions, independence, and convexity. *Synthese*, *186*, 577–600.
Cozman, F., & Walley, P. (2001). Graphoid properties of epistemic irrelevance and independence. In *Proceedings of the 2nd international symposium on imprecise probabilities and their applications* (pp. 112–121).
Danks, D. (2014). *Unifying the mind: Cognitive representations as graphical models*. Cambridge, MA: MIT Press.
De Bock, J. (2017). Credal networks under epistemic irrelevance. *International Journal of Approximate Reasoning*, *85*, 107–138.
De Bock, J., & de Cooman, G. (2015). Credal networks under epistemic irrelevance: The sets of desirable gambles approach. *International Journal of Approximate Reasoning*, *56*, 178–207.
De Bock, J., & de Cooman, G. (2019). Interpreting, axiomatising and representing coherent choice functions in terms of desirability. In *Proceedings of machine learning research (ISIPTA 2019)* (Vol. 103, pp. 125–134).
Elga, A. (2010). Subjective probabilities should be sharp. *Philosophers' Imprint*, *10*(5), 1–11.
Gärdenfors, P., & Sahlin, N. (1982). Unreliable probabilities, risk taking, and decision making. *Synthese*, *53*, 361–386.
Glymour, C. (2001). *The mind's arrows: Bayes nets and graphical causal models in psychology*. Cambridge, MA: MIT Press.
Glymour, C., & Danks, D. (2007). Reasons as causes in Bayesian epistemology. *Journal of Philosophy*, *104*(9), 464–474.
Gopnik, A., & Schulz, L. (2007). *Causal learning: Psychology, philosophy, and computation*. New York: Oxford University Press.
Hitchcock, C. (1996). Causal decision theory and decision-theoretic causation. *Noûs*, *30*(4), 508–526.
Jeffrey, R. (1965). *The logic of decision*. New York: McGraw Hill.
Jeffrey, R. (1983). Bayesianism with a human face. In J. Earman (Ed.), *Testing scientific theories* (pp. 133–156). Minneapolis: University of Minnesota Press.
Joyce, J. (1999). *The foundations of causal decision theory*. Cambridge: Cambridge University Press.
Joyce, J. (2005). How degrees of belief reflect evidence. *Philosophical Perspectives*, *19*, 153–179.
Joyce, J. (2010). A defense of imprecise credences in inference and decision making. *Philosophical Perspectives*, *24*, 281–323.
Kadane, J. B., Schervish, M. J., & Seidenfeld, T. (2004). A Rubinesque theory of decision. *IMS Lecture Notes Monograph*, *45*, 1–11.
Kaplan, M. (1996). *Decision theory as philosophy*. Cambridge: Cambridge University Press.
Kyburg, H. (1983). *Epistemology and inference*. Minneapolis: University of Minnesota Press.
Levi, I. (1974). On indeterminate probabilities. *Journal of Philosophy*, *71*, 397–418.
Levi, I. (1980). *The enterprise of knowledge*. Cambridge, MA: MIT Press.
Meek, C., & Glymour, C. (1994). Conditioning and intervening. *British Journal for the Philosophy of Science*, *45*, 1001–1021.
Menzies, P., & Price, H. (1993). Causation as a secondary quality. *British Journal for the Philosophy of Science*, *44*, 187–203.
Pearl, J. (1988). *Probabilistic reasoning in intelligent systems*. San Mateo, CA: Morgan Kaufmann.
Pearl, J. (2009). *Causality: Models, reasoning, and inference* (2nd ed.). Cambridge: Cambridge University Press.
Price, H. (1991). Agency and probabilistic causality. *British Journal for the Philosophy of Science*, *42*, 157–176.
Seidenfeld, T., Schervish, M. J., & Kadane, J. B. (1995). A representation of partially ordered preferences. *Annals of Statistics*, *23*, 2168–2217.
Seidenfeld, T., Schervish, M. J., & Kadane, J. B. (2010). Coherent choice functions under uncertainty. *Synthese*, *172*(1), 157–176.
Sen, A. (1977). Social choice theory: A re-examination. *Econometrica*, *45*, 53–89.
Spirtes, P., Glymour, G., & Scheines, R. (2000). *Causation, prediction and search* (2nd ed.). Cambridge, MA: MIT Press.

Spohn, W. (2000). Bayesian nets are all there is to causal dependence. In M. C. Galavotti, et al. (Eds.), *Stochastic dependence and causality* (pp. 157–172). Stanford: CSLI Publications.

Stern, R. (2017). Interventionist decision theory. *Synthese*, *194*(10), 4133–4153.

Sturgeon, S. (2008). Reason and the grain of belief. *Noûs*, *42*, 139–165.

Van Camp, A. (2018). *Choice functions as a tool to model uncertainty*. PhD thesis, Faculty of Engineering and Architecture, Ghent University.

Van Camp, A., de Cooman, G., & Miranda, E. (2018). Lexicographic choice functions. *International Journal of Approximate Reasoning*, *92*, 97–119.

von Neumann, J., & Morgenstern, O. (1947). *Theory of games and economic behavior*. Princeton: Princeton University Press.

Walley, P. (1991). *Statistical reasoning with imprecise probabilities*. New York: Chapman and Hall.

Williamson, J. (2001). Foundations for Bayesian networks. In D. Corfield & J. Williamson (Eds.), *Foundations of Bayesianism* (pp. 75–115). Berlin: Springer.

Woodward, J. (2003). *Making things happen: A theory of causal explanation*. Oxford: Oxford University Press.

Yule, G. U. (1903). Notes on the theory of association of attributes in statistics. *Biometrika*, *2*, 121–34.

Zhang, J., & Spirtes, P. (2011). Intervention, determinism, and the causal minimality condition. *Synthese*, *182*, 335–347.

The manufacturer's authorised representative in the EU is Springer Nature Customer Service Centre GmbH, Europaplatz 3, 69115 Heidelberg, Germany. If you have any concerns regarding our products, please contact ProductSafety@springernature.com

Printed and bound by CPI Group (UK) Ltd, Croydon, CR0 4YY
28/09/2026
02224088-0003